FREEDOM'S EDGE:
THE COMPUTER THREAT TO SOCIETY

FREEDOM'S EDGE:
THE COMPUTER THREAT TO SOCIETY

Illustrations by
Will Eisner

MILTON R. WESSEL

 ADDISON-WESLEY PUBLISHING COMPANY
Reading, Massachusetts
Menlo Park, California
London · Amsterdam
Don Mills, Ontario
Sydney

Third printing, August 1975

Copyright © 1974 by Addison-Wesley Publishing Company, Inc. Philippines copyright 1974 by Addison-Wesley Publishing Company, Inc.

All rights reserved. No part of this publication may be reproduced, stored in a retrieval system, or transmitted, in any form or by any means, electronic, mechanical, photocopying, recording, or otherwise, without the prior written permission of the publisher. Printed in the United States of America. Published simultaneously in Canada. Library of Congress Catalog Card No. 74-4717.

ISBN 0-201-08543-7
BCDEFGHIJ-AL-79

To Joan, who makes
all things possible

FOREWORD

In this time of rapid inflation, when prices seem to increase almost daily, the price of one commodity—computing (information-processing) power—has been declining rapidly. Today, there is a small computer which retails for less than $1000. Twenty-five years ago the equivalent computational power would have cost several million dollars.

This decrease in price has been matched by a decrease in size. I can hold today's small computer in my hand, whereas the equivalent computer of 25 years ago filled a large room and needed another room full of air-conditioning and power-stabilization equipment. (However, not all of today's computers are small enough to be hand-held, because most are much more powerful and faster than the hand-held model I refer to, and more important, they have printers, card readers, and various other peripheral devices attached to them—facilities which the hand-held model does not have.)

Impressive as the miniaturization may be, it is the rapid decline in cost of computing power which accounts for its pervasiveness and impact on society, and those costs continue to plummet. It is reasonable to predict that by the time Orwell's 1984 is upon us (take that statement either way you like), the cost of computing power, compared with today's cost, will have declined by a factor of at least 100, or perhaps 1000 or more.

To be more precise, I must point out that these very rapid declines are associated with the electronic portions of computers. Getting a computer to do useful work also requires human beings whose wages have been increasing, although probably not as fast as their productivity. In any event the total cost, including the labor, of doing a given information-processing task has declined rapidly. Thus, organizations have larger and larger budgets for information-processing as more and more applications become economically feasible as the costs decline. Also, government and industry store in their computers more and more information about individuals and organizations.

It has often been pointed out that when something changes by a factor of ten (often referred to as an order of magnitude), it has profound effects on society. For example, a person can walk at about 4 miles per hour, travel by automobile at something like 40 mph, or go by jet aircraft at roughly 400 mph. The Apollo astronauts averaged about 4000 mph in their trips to the moon. Each of these differs from the previous by a factor of ten—4–40–400–4000—from walking to space travel in three orders of magnitude. The increase to 40 mph made possible the flight to the suburbs and drastically changed our mores. The jet aircraft has shrunk our world immeasureably. The impact of space travel is just beginning to be felt.

In the computer field even greater orders of magnitude changes are taking place in a short period of time. I have said that by 1984 the cost of computing power will decline by at least another factor of 100. Suppose it were reasonable to predict that by 1984 automobiles or houses would cost one one-hundreth of today's cost (thus a new Cadillac would cost less than $100)—or to predict that energy would cost one hundred times as much. All these involve a change of two orders of magnitude, and we would expect all of them to have a profound impact on our way of life. Thus we cannot avoid the conclusion that a change of two orders of magnitude in computer costs will alter our way of life appreciably, probably in some ways we cannot foresee today.

It is the nature of many of the problems facing our society today that several specialities or disciplines are involved. Thus, air pollution involves chemistry, meteorology, physics, engineering, economics, and other topics. The impact of computers on society involves at least computer science, law, psychology, sociology, economics, and political science. As a writer on this topic, Milton Wessel is a bit unusual in that he was educated as a lawyer and has come to know the computer field through many years as legal counsel to AFIPS (American Federation of Information Processing Societies, an association of the professional societies in the computer field) and to ADAPSO (Association of Data Processing Service Organizations, a trade association). As a lawyer, Mr. Wessel brings to his subject a perspective which most writers on "computers and society" do not have. What he has to say deserves the attention of every citizen.

This book does not emphasize the many benefits that computers have brought or will bring to society. Rather, it focuses on some of the disadvantages and potential dangers. Computer personnel may worry about the balance, between pluses and minuses, in this book for fear that modern-day Luddites may be incited to destroy their machines. That is not Mr. Wessel's intent; rather, he would have us understand and face up to the problems so that society may realize the full potential of the computer without undesirable side effects.

Paul Armer
*Center for Advanced Study in the Behavioral Sciences,
Stanford, California*

PREFACE

A communications medium transmits messages. It also may affect the message itself.

A computer system processes data. It also may affect the data itself. It is the theme of this book that when the computer's impact on the data is great enough, it changes the environment in which we live.

These chapters consider some of the ways in which the computer is already changing our environments and our lives, or soon will be. For every benefit the computer gives us, most of which are easily and quickly recognized and appreciated, lurking beneath the surface there is often an accompanying insidious and inadequately understood or unrecognized detriment to something we cherish.

Especially in America we have been prompt to accept the benefits offered by the computer; the world is now following rapidly and is no longer far behind. The evidence of this is already all around us, from the omnipresent MICR-encoded bank check and punch card utility bill ("don't fold, staple, or mutilate") to the airline reservation and brokerage quotation terminal. We simply couldn't live as we do today without the computer.

But with the single exception of the "privacy" issue, we have hardly begun even to recognize the detriments, much less analyze or deal with them. With only a few minor exceptions our universities, government agencies, foundations, and intellectual spokesmen have been almost notoriously silent. The inertia and apathy incident to any development so new and different may be one reason for this. However, I believe the key reason is a deep-seated fear of displaying ignorance of the new technology, nurtured by the tendency of so many computer scientists to use an almost incomprehensible "computerese."

I do not and cannot pretend to explore in these pages *all* of the societal problems the computer is creating. The presently inadequate consideration of computer societal impact is an important reason why many—perhaps most—such problems are still unidentified. (A second reason, of course, is that we simply do not yet have all the information necessary to anticipate what will come.) I suspect that what is described represents only the extreme tip of the iceberg. For precisely the same reason, those few solutions which are suggested are advanced cautiously, tentatively, and for discussion rather than adoption.

I know that this work is filled with my own personal social and political views and prejudices, positions which I have not tried to support or even argue. Despite my strong convictions regarding privacy, I appreciate that some people are willing to volunteer the most intimate private information about themselves without duress, even when they know it can be used against them. Others will also differ with my beliefs that individual freedom generally should be accorded a higher priority in the scheme of things than the services provided by the computer, or that strengthening the position of the establishment is usually undesirable. It would detract from my theme to argue these strictly collateral issues. The effort will have well filled its purpose if it serves only to generate interest and concern, and thereby stimulates the necessary analysis, debate, and action before it is too late; before the computer has become the master of our fates in a fashion even beyond the dreams of science-fiction writers.

※ ※ ※

The sometimes simplistic presentations of this book may concern some of my friends and colleagues, especially those in the computer industry. Without attributing any responsibility to them, I

must express my deep appreciation for the help they have given. To select a few is always dangerous, but I cannot go to print publicly without acknowledging the tremendous guidance, support, and help I have been given over the years by Bruce Gilchrist, Paul Armer, Jerry Dreyer, and Bernie Goldstein. I must also express special appreciation for the research assistance and advice of Prof. dr. Stanislaw J. Soltysinski of the University of Poznan, to Phil Dorn for his straightforward critiques and enormous output, and to my incredible secretary, Susanna Levy.

M.R.W.

New York
May 1974

CONTENTS

1
THE COMPUTER DILEMMA 1

2
A COMPUTER UTILITY BILL OF RIGHTS 9

3
**FAIR ACCESS TO THE CHECKLESS/
CASHLESS SOCIETY** 17

4
**THE DATA BANK—CIVIL RIGHTS AND
CIVIL WRONGS** 24
    Chilling Effect of Data Bank 30
    The Government and the Governed 32
    Inadequacy of Present Data-Bank Proposals . . . 34
    Accurate vs. Inaccurate Data 35
    Data-In-Gross vs. Transactional Data 37
    Voluntary vs. Surreptitiously Obtained Data . . . 39
    Objective vs. Subjective Data 40
    Private vs. Public Data Banks 42
    Solutions 45
    Current Legal Guidelines 50

    The Data-Bank Public Impact Statement **58**
    Privacy Commission **60**
    Swedish Data-Inspection Board **63**

5
POINT-OF-SALE MARKETING AND FREE COMPETITION **69**

6
SECURING THE ESTABLISHMENT **83**

7
UNTYING THE COMPUTER GRID **91**
    Computer-Related Crime **94**

8
THE VITAL HUMAN FACTOR **97**
    Human Response **97**
    Human Obsolescence **108**
    Computer Intellectual Property **109**
    Man's Image of Man **112**

9
CONTROLLING THE COMPUTER **119**
EPILOGUE **131**
APPENDIX **135**
    The Ten Commandments of Computer Usage **135**

Whenever we introduced a new technological convenience, we had to examine its place in the life of our institution most carefully. The advantages we could enjoy from any new machine were always quite obvious; the bondage we entered by using it was much harder to assess, and much more elusive. Often we were unaware of its negative effects until after long use. By then we had come to rely on it so much, that small disadvantages that came with the use of any one contrivance seemed too trivial to warrant giving it up, or to change the pattern we had fallen into by using it. Nevertheless, when combined with the many other devices, it added up to a significant and undesirable change in the pattern of our life and work.

This is what I mean by "seduction." The advantages of the machines are so obvious and so desirable, that we tend to become, small step by small step, seduced into ignoring the price we pay for their unthinking use. The emphasis here is on *unthinking use*, because they all have their good uses. But the most careful thinking and planning is needed to enjoy the good use of any technical contrivance without paying a price for it in human freedom.

> Bruno Bettelheim (writing of his work with psychotic children), *The Informed Heart,* New York: Avon Books, 1960. Reprinted by permission.

1 THE COMPUTER DILEMMA

It was 1953—the height of the McCarthy era. The American atmosphere was charged with suspicion, distrust, and hatred. Accusations of being a "fellow traveler" or a "Communist front," based on the flimsiest of evidence or contrived evidence or even none at all, were enough to destroy a career and a lifetime of effort.

I was an Assistant United States Attorney for the Southern District of New York, air bound for Washington. Mine was a particularly sensitive prosecutor's office, for we were handling many of the important Smith Act and other subversion cases. The 1951 *Rosenberg* convictions in our District, the first to end with civilians put to death for espionage by order of a nonmilitary court, were still making front-page headlines. Roy Cohn, notorious for frequently out-McCarthying the Wisconsin Senator, had only recently left our office to join the McCarthy Committee as Chief Counsel, and had many friends among us. The Committee had held widely publicized hearings in Courtroom 110 at the United States Court House in New York City's Foley Square where we were located, attended by television cameras, publicity seekers, and much excitement. I recall going downstairs one day to watch, being much embarrassed but drawn by the circus character of the supposedly quasijudicial proceeding, and hiding in the background lest I be photographed, or someone see and greet me or call me onstage. I left quietly and very quickly, despite my great interest in the spectacle.

Already there had been leaks of classified information to the Committee. All of us were intensely security conscious. Little wonder that when the stewardess handed me a copy of a current *Life* magazine, the cover of which announced a critical examination of Senator McCarthy and his activities, I flipped quickly to the article and began to read.

My memory of that trip almost two decades ago is otherwise clouded, but I vividly remember sitting on the right-hand side of the plane, halfway back in an aisle seat in the coach section. A few seconds after turning to the article I became conscious of a man sitting next to me in the window seat, a man whose face remains a blur which I may never have seen. I reflected for no more than a moment about who my fellow passenger might be—what he might think about my reading the article and the reaction on my face, whether I smiled or looked glum, how quickly I read, what he might say and to whom he might report—and then almost immediately I flipped on through the magazine to some other section.

Later I purchased another copy of the magazine at a newsstand and read it secretly in my hotel room—a little ashamed of myself.

* * *

It was 1958. I had resigned as an Assistant United States Attorney, returned to private practice, and was now back in government as a Special Assistant to the United States Attorney General. I was the chief organized crime prosecutor in the United States, probing the November 1957 Apalachin "crime convention" in an attempt to find out what was wrong with organized crime enforcement and recommend improvements. Inevitably the assignment had brought me into a head-on clash with J. Edgar Hoover, then the publicly much venerated Director of the Federal Bureau of Investigation.

From the outset Hoover considered me an interloper in areas he had always controlled. He used every tool in his book to block submission of my recommendations for improvement, ideas which he regarded as highly critical of the FBI even though they mentioned the FBI by name only once. Although unsuccessful in this, he did prevent their adoption and implementation. His antagonism ultimately became so great that he sometimes lost his usual "cool" and, in public, his appearance of impartiality and aloofness from the fray. He attacked directly in Congressional testimony and public statements, rather than indirectly through his numerous friends in Congress, the press, and elsewhere. (The indirect approach had been his common practice, for he was a master of publicity and public relations as well as of political intrigue.)

Technically at least, the FBI was a part of the same Department of Justice and responsible to the same Attorney General as were my unit and I. As the Department's investigative arm, the FBI accumulated files on all Department personnel and made them available to the Attorney General and his assistants as it considered appropriate. Security was again a touchstone of our operation, and from whatever source, this kind of information concerning me and my assistants was evaluated carefully and promptly. I studied the data on each of the key regional chiefs, and they in turn examined the reports with respect to their subordinates. No one was permitted to see his own file, and Dick Ogilvie, then Chief of the Midwestern Region of our Special Group on Organized Crime and later Governor of Illinois, reviewed the information about me.

Again, time has blurred the details, but I do clearly recall sitting in Assistant Deputy Attorney General John Sheneman's office while Dick and I inspected some files. Dick was looking at one of the early FBI reports on me. At one point he said, "I see your father signed the Stockholm Peace Pledge in 1945."

Even as late as 1958, subscription to the Stockholm Peace Pledge was accepted by many as evidencing subversive or at least "fellow traveler" tendencies, for among its sponsors were organizations identified as "Communist fronts" on the Attorney General's list of subversive organizations. The pledge itself was a document circulated world-wide during World War II while America was actively allied with Russia, expressing innocuous-

4   The Computer Dilemma

sounding platitudes adding up to a plea that the nations of the world unite to stop the brutality and destruction of war.

My father's three sons had all been overseas during the War, serving in the European Theatre of Operations; one had crossed the Channel to France the day after D-Day, and for a period was missing in the Battle of the Bulge: the second, an Infantry Private, had come to Europe through North Africa, Sicily, and on through the Italian boot, much of the time carrying a part of a mortar on his back. I had been in England with the 8th and 9th Air Forces. My father loved his sons deeply, and his concern for their safety was great. But none of the latter information was in the investigative report.

I looked at Dick Ogilvie after he revealed this perhaps intentionally selected bit of "derogatory" information in my file, and he looked at me as if to say, "Can this be happening here?" But I still remember being careful to find a later opportunity, without appearing too obvious, to tell him that my father was a lawyer and a conservative, and had been a registered Republican all his life.

By this time I had had enough further experience with security matters and the use of facts to override any shame at feeling a need to protect the record, even though I knew my father and me to be innocent of any of the adverse implications of the fact revealed.

* * *

Neither of these two isolated episodes has anything to do with the computer; each took place years before it became a significant factor in our society. Instead they reflect the deep concern of one individual—trained as a trial lawyer and prosecutor in the marshaling, selection, and use of facts—about the possible adverse use against him of an isolated fact which might be brought to light years later in an unrelated context; about how and why one might measure and adjust one's conduct accordingly; and about what this means to freedom.

Only a person in a most sensitive position and trained in the use of facts would have reacted as I did to these events. Even today details such as how a man traveling alone on an airplane looks while reading a magazine article, or whether a father was one of several million signatories to a world plea for peace thirteen years before, can be processed only for a miniscule fraction of society. Tomorrow, however, the computer will make possible the accumulation and instant accessibility of far more such minutiae, reaching into our most intimate moments. The implications are both tremendous and terrifying.

Falstaff may have been correct that discretion is the better part of valor, but I was not pleased with my reaction or conduct in either of these episodes. What happened surprised (and disappointed) me at the time; it would not surprise me now. I fear that in another decade it may not surprise anyone.

This is the dilemma of the computer, which is apparent whenever one probes its impact. Every exciting new benefit the computer offers, such as the analyzing of a myriad of isolated facts to aid in criminal prosecution, is accompanied by a concomitant risk often obscured by the promise of the benefit.

Even more, I fear the other not-yet-apparent adverse consequences of the computer dilemma. The computer has a tremendous capacity to retain a staggering volume of information, to process and control it almost instantaneously, and to make it available in virtually unlimited ways. Unless we recognize and understand its adverse consequences, however, and do something about them before it is too late, the computer may create a way of life very different from anything we have had before—one in which freedom as we know it (or at least as the more fortunate among us do) simply cannot exist. For example:

> The computer makes possible incredible oral, visual and written communications capabilities which can inform and educate the masses instantly. At the same time, however, this can vest such huge control in the hands of one or a few persons, as to threaten the rise of new Hitlers and Stalins wielding powers far greater than their predecessors.
>
> The computer makes possible instant mass voting and opinion polls which can mean unparalleled responsiveness to the

wishes of the public. This identical capability, however, also creates the potential for destruction of traditional limitations on majority control.

The computer makes possible our future "checkless/cashless" banking structure which can process and help produce enormous material wealth. This same structure, however, can further isolate and make pariahs of the impoverished, the unwanted, and the deviates.

The computer makes possible marketing and manufacturing point of sale and process controls which can effect labor savings reducing the work week to almost any minimum. This very same efficiency can also mean undesirable vertical integration and oligopoly or monopoly in a few resources of supply, destroying freedom of economic opportunity.

Human technological specialization, to this point an apparently unavoidable feature of the computer era, can create a set of persons controlling the productive segment of society which is not equipped to deal with traditional values. It can make it increasingly difficult for democratically selected lay leaders to function intelligently. It can result in human technological obsolescence at an early age, developing a new class of unwanted which will pose a far more difficult problem than the science of geriatrics has yet seen. To some extent it already has. I know some members, young in years but old in despair.

The computer's threat to privacy is already documented and recognized in government and private circles, and at least some corrective action seems underway; the degree of attention focused on that area may indeed have contributed to obfuscation of more serious risks. It may well be that some of these other computer-associated developments will ultimately turn out to have far more significant adverse impact on our lives than the vast storehouses of information to be built up in credit files and government dossiers, about which so much has been said.

Noncomputer people are most reluctant to disclose their ignorance of the computer's special esoteric language and opera-

tions. As a result, there is precious little discussion and debate about what is happening and its consequences within other disciplines in the universities and professional associations and among the public generally. Before it is too late, our society requires intensive, nontechnical, free-swinging interdisciplinary recognition and discussion of these matters. Such a discussion should produce the information enabling each of us to decide for himself what measures are required for the public good. It would thus make it possible for the risk/benefit issues posed by the computer dilemma to be resolved by everyman, not by technocrats or the computer itself, so that we will all be masters of the new technology rather than its servants. If and when we are in effective control, the computer will be neither a menace nor a messiah, but simply an extraordinarily valuable machine to be used to help us move forward to a better life.

2 A COMPUTER UTILITY BILL OF RIGHTS

The "computer utility" sounds like a phrase descriptive of that day far in the future when each of us will be able to plug into computer power, just as today we plug into and turn on electric power from the electric utility, telephone service from the telephone company and water from the water company. Although the analogies may be extreme, the day is *not* far off in the future. In a somewhat rudimentary form it is here right now.

The air traveler knows how absolutely dependent modern air travel and service is on the computer. The mere statement, "Sorry, SABRE's down," from an American Airlines agent is enough to mean a change to another airline or postponement of plans until service personnel are able to get SABRE, American Airlines' computerized reservation service, back in operation. And the investor equally knows how dependent his broker is on QUOTRON or other computerized brokerage information service, for information on current stock market prices or a company's financial history. SABRE and QUOTRON are two existing operating examples of special purpose application precursors to the full-fledged "computer utility," as that term is used in this book.

"Computer utility" is a phrase employed rather widely in the computer industry, but not much elsewhere. It has not yet come to have any precise meaning. It is employed to describe use of a distant or remote computer by highly sophisticated computer professionals who supply substantially all their own programs ("programs" are part of the instructions which make the computer work) and require little more of the utility than raw computer power. The term is also used to describe such use of a computer by clerks, secretaries, and various lay personnel, who tap into the computer through a terminal something like a typewriter. The latter users need know nothing more about the computer's operation than they do about that of the radio or television set they turn on in similar fashion. It is this second category of applications by noncomputer people which is pertinent here and is the sense in which I use the term "computer utility" in this book. This is the use which has the most direct and personal impact on our lives. It is the use which the public is almost by definition least equipped to control. The word "utility" for this reason may be especially appropriate, because it has a second connotation apart from usefulness: it implies governmental regulation. Although computer professionals using the phrase "computer utility" by and large have not yet overtly proposed such control, subconsciously they may be expressing that intention.

The computer utility of the future will reach directly into our homes via telephone lines, cable television, or even wireless communication. It may appear to us on our television screen, through a typewriter or printing terminal, or by voice. It will offer us a host of information services with which we can interact, permitting us to shop at our supermarket, participate in a classroom-type learning experience, or voice our opinions in a market research survey or to our Congressman and Senator. This is not science fiction. Although these categories of computer utility services are not yet being offered, they are well within current technological capabilities. The Federal Communications Commission has in fact already published rules requiring that this two-way interactive capacity be built into certain types of cable communications systems as a prerequisite to licensing. Whether it is five, ten, or more years away is little more than a matter of economics at this point.

The value of these kinds of services staggers the imagination. Mass adult public education, as well as training of the sick and bedridden, becomes a reality. The viewer can sit at his television screen, watch a lecture or visual presentation, and be tested on his comprehension. Computer *aided instruction* (CAI), in which the student communicates back and forth with his computer-teacher, is no further away than the living room. The chore of marketing for the working wife or husband can be far less burdensome; and for the selective buyer, the opportunity to shop carefully, to compare prices and qualities (and thereby maximize purchasing power), can be a reality. Opinion surveys no longer need be limited to the observation of selected categories and samples, but can obtain responses from millions of homes instantaneously, with analyses possible by geographic area, sex, age, or just about any other quality selected. Elected representatives need no longer guess the desires of their constituencies —they can obtain prompt statements of attitudes and preferences from any who care to comment. The computer is not the universal panacea in any respect, and there is a limit to the extent to which even the best-informed public opinion should be permitted to control complex government policies. But surely in this fashion more effective participatory democracy can be facilitated.

But—and there is a "but" here as there is with every other promising computer application—these benefits do not come without risks and costs. Although the computer may not be watching each of us from its television terminal in "Big Brother" fashion, its influence can be just as antisocial. We need a computer utility "bill of rights" to limit these antisocial effects. Steps must be taken to ensure that access to any large national computer utility system is open and fairly distributed (*fair access*). The information the utility furnishes to the user must be fully and fairly disclosed so that the user's responsive action can be intelligent (*fair disclosure*). Overall, use of the computer utility must serve the public interest (*fair use*).

The retailer with direct access into your home to offer his wares may be delighted with his sales. But what about the hundreds or thousands of others competing for your time and money? Cable television offers far greater channel potential for carrying computerized data than wireless UHF or VHF television, and electro-optics is already on the horizon, permitting the transmission of 100 billion bits per second ("bits" are the smallest units of computerized data), ten thousand times more than cable. Shortwave satellite transmission may add a further dimension to the channel spectrum. At least for the near future, however, it is unlikely that there will be enough capacity for every supplier to interact with every customer and in any event, there is a very real limit to the number of cable carriers which can have access to any one home or any single area. Because of such considerations, many computer utilities will be just as much natural monopolies as the electric utility and the telephone and water companies referred to at the outset of this chapter. One is reminded of those several hours during the 1972 Munich Olympic games when the COMSAT satellite's time was fully allocated and prevented transmission of news of the Arab kidnapping and slaughter of the Israeli athletes. The computer utility threatens similarly to restrain competition and to foster monopoly both in trade and ideas.

The highest courts of Massachusetts and Florida have disagreed on the validity of state statutes that require newspapers to publish the replies of political candidates who are criticized either editorially or in the newspapers' advertising, and the U.S. Supreme Court is reviewing the issue. It may also be that the present strict free "equal-time" broadcasting rule will be modified or even abolished. But the Fairness Doctrine will still require that contrasting viewpoints be aired on controversies of public importance. Candidates in political campaigns will still be entitled to fair use of at least the radio and television media. The general rule of present antitrust law is that a monopolist, unlike a normal businessman, cannot refuse to deal in order to maintain or extend his commercial position. A related formulation of substantially the same principle is the rule that those controlling use of an essential facility must grant access on reasonable and nondiscriminatory terms to all in the trade. Rights to a wide variety of other benefits cannot be denied today on the basis of race, sex, or certain other types of characteristics. As the January 14, 1974, report of the U.S. Cabinet Committee on Cable Communications concluded in another connection:

> If cable is to become a constructive force in our national life, it must be open to all Americans. There must be relatively

easy access at one end of the cable for those who wish to promote their ideas, state their views, or sell their goods and services; and at the other end, the consumer must have a meaningful freedom of choice to select from among a diverse range of cable programming and services.

A similar doctrine of fair access, allocating home-interconnect capability equitably among competing vendors and other users, seems essential when the computer utility is with us. All of this leads to:

Computer Utility Rule 1: Access to a computer utility system shall not unreasonably be withheld.

❊ ❊ ❊

The Congressman seeking to make participatory democracy a reality by polling his constituency may have good intentions; but unless he is an expert in opinion surveys, he may create an apparent public attitude which does not exist and may be in fact contrary to what the public would want were it informed. His vote may even be contrary to the action he would have taken had he not been misled by the apparent results. This is because the phrasing of a question itself may determine the answer. "Shall we surrender to the Communists in Vietnam?" and "Shall we make peace in Vietnam?" are very different questions which might get different answers even from the same person. The computer utility threatens to encourage political, economic and social action which *appears* to be consistent with the belief and views of the public and thereby to foster participatory democracy, but which is *actually* inconsistent or even in conflict with the public's true desires.

The SEC's full disclosure principle required in connection with the sale of securities and for stockholder proxy statements assures at least an opportunity to understand the issues. An analogous doctrine of fair disclosure also seems appropriate in connection with any large-scale opinion or voting computer application where the result is to be given meaningful use. This leads us to:

Computer Utility Rule 2: The information disclosed by a computer utility system seeking response must be such as to permit the respondent to provide an intelligent answer.

* * *

The teacher developing a curriculum for the local school is subject to all the controls of his school board, superintendent, principal, and department head. Yet the teacher developing and offering a CAI application for the education of millions of adults is pretty much free to teach what he wants, although it may have far more serious adverse impact. The computer utility thus threatens to interfere with proper standards of public information and education.

Television limits or prohibits cigarette and liquor advertising, and certain excesses of slang and sex. A similar doctrine of fair use of the computer utility is also necessary to control the power which this new weapon threatens to have over our minds. This brings us to:

Computer Utility Rule 3: The information furnished by a computer utility system must be such as to serve the public interest.

* * *

The doctrines of fair access, fair disclosure and fair use outline concepts in the broadest and most general terms. This seems necessary at this point in the computer age, because we do not yet have all the details of how these utility systems will operate and be applied, nor much experience with their use. We must avoid the sort of rigorous and stultifying rule application that has resulted from the equal-time broadcasting rule in political campaigns. And we must be especially careful not to impinge on freedom of speech and other civil liberties. But difficulty and uncertainty must not stand in the way of beginning, and clearly there is ample precedent for dealing with these kinds of general rules.

The primary building block of the common law tort of negligence is the concept of the "reasonable man." A person who unintentionally causes harm to another is not responsible if he acted at least as carefully as the reasonable man would have acted under the circumstances; he is liable for damages if his performance did not measure up to the manner in which the reasonable man would have conducted himself. The courts have developed workable tools to decide how a reasonable man would act under the most extreme or novel conditions, and even as the measure of the reasonable man has changed with the times. A reasonable man in the eighteenth century might have cauterized a wound following injury; in the twentieth century he might telephone a doctor or call for an ambulance; in the twenty-first he might consult the computer.

Constitutions and statutes are frequently deliberately drafted in this same kind of broad general language, to permit subsequent construction and application to reflect experience and changing social and economic patterns. Just as with the "reasonable man" of common law, courts and administrative agencies have defined and given changing content over the years to constitutional phrases such as *due process* and *equal protection,* and to statutory terms such as *restraint of trade.* They should not have any greater difficulty with the computer utility bill-of-rights concepts of fair access, fair disclosure, and fair use.

Realistically and flexibly defined and applied, these doctrines can channel the computer utility's power into directions which will serve our needs and enrich our lives. Without them, we may find that the computer utility has created material wealth but destroyed the intangibles which make life worth living.

3 FAIR ACCESS TO THE CHECKLESS/CASHLESS SOCIETY

The gasoline station at which my family deals in our home town of White Plains, New York, will not sell gasoline to us at night, unless we use a credit card or have the exact amount of cash to be dropped into a heavy, theft-proof locked box. A number of the fine little shops in the high-rent midtown area along New York's Madison Avenue retain only a minimum of cash on hand and keep their doors locked and secure until they have a chance to glance at the customer to see whether he looks like a thief. To these people, to the New York bus driver (who carries no change and accepts fares only when dropped into a locked box), to the taxi driver (who by law is not required to change more than a five-dollar bill), and to a burgeoning number of others in our high-crime metropolitan areas, cash constitutes a serious threat to survival.

Bank checks are one substitute for cash. But they are not legal tender; they represent a credit risk to the vendor and are an acceptable alternative to the already ubiquitous credit card only on limited occasions and in limited respects. Moreover, even with *m*agnetic *i*nk *c*haracter *r*ecognition (MICR)—the little magnetic numbers on the bottom of virtually all American checks today, which permit checks to be processed electronically—we are rapidly approaching the limit to the number of separate paper checks our banks can handle. The Federal Reserve Board estimates that about 26 billion separate bank checks were processed during 1973, in most cases efficiently and effectively. If present growth rates persist, this will double during the next decade and something new will be required to handle the volume of paper. The banking fraternity has been especially diligent and ingenious in its planning, and already has that "something new" on the drawing boards and in test. It is the checkless/cashless society—a catchy phrase for what is really intended to be an economy utilizing fewer checks and less cash.

Although there are major differences, at first glance the checkless/cashless society seems to be little more than an extension of the credit card to permit a customer's bank account to be charged electronically at the moment of purchase, and the vendor's account credited without the usual paper check passing from customer to vendor to vendor's bank to clearing house to customer's bank. Other phrases sometimes used to describe all this are "electronic funds transfer system" (EFTS) and the "automated (or 'electronic') payments mechanism."

Bank credit cards such as Master Charge and BankAmericard are not just alternatives to American Express and Diner's Club cards, but forerunners of this checkless/cashless society. So is the "instant money" service now offered by a few banks, which permits a card holder to insert his bank card in a small type of computer and thereby obtain $25 or $50 at night or on a weekend or holiday when the banks are closed. All the holder has to do is punch out on a computer terminal a secret magnetic number encoded on a strip on the back of his card. The "credit card" thus becomes a "debit card" as well, which is yet another step towards automation of the economy. The increasing number of plans by which banks deposit salaries to employees' accounts without paycheck, or pay regular rent, insurance, and other bills for customers, are even more common examples of these advance agents of the checkless/cashless society.

Bankers are meticulous and precise people, and there is not yet any general consensus as to exactly how the checkless/cashless society will work, or how universal its application will be and how soon. Even the technicians are not agreed as to whether the identification will be by voice, signature, fingerprint, or secret

Fair Access to the Checkless/Cashless Society

number. These are details which admittedly are complicated by the difficulties of finding foolproof methods in today's rip-off society. The economics of simple efficiency, plus the rising crime problem and limits to check handling capability, suggest that the following is an accurate general description of how it will work.

You will have a bank credit/debit card, very much like the one you have today, with a magnetic strip on the back carrying a secret number or other unique means of identification, exclusively yours. Suppose you live in Los Angeles and are visiting New York with your wife on a long-planned twenty-fifth wedding anniversary trip. Shopping is very much a part of your vacation, and you wander in and out of some of the finer Fifth Avenue shops. Finally she spots just the thing—crystal glassware for bar and table at Tiffany & Co., price $665, tax included.

You hand the sales person your card, which he inserts in a small holder very much like today's credit card imprinters, but which is connected to his touch-tone telephone (another harbinger of the computer age already available throughout the country). He taps a few numbers on the telephone, and a green light signals that the card was properly issued and has not been reported lost or stolen. He hands you the phone and you tap out your secret identifying number. The green light begins to blink, confirming that you are (or at least appear to be) you.

"Cash or credit?" inquires the sales person. "Cash, please," you reply, "if my balance is sufficient." The sales person taps out some more numbers, but this time a red light appears. "Sorry," he says, "your account's too low. Shall I try credit?" You nod affirmatively, and he taps out some more numbers, but again the red light appears. You remember that your bank line of credit is only $1500, and you've already borrowed a great deal on this trip. "Try $400 cash and $265 credit," you suggest. He does, the green light blinks happily twice, once for the cash and once for the credit transaction, and the sale is completed after you sign a sales slip confirming the purchase and he gives you a receipt.

At the instant of the sale, your Los Angeles bank balance was reduced by $400, and your loan account (at 18 percent interest, annual rate) was increased by $265. Tiffany's bank balance was increased by $665, its inventory of crystal glassware reduced by the number you purchased, its sales person's commission account adjusted, and a host of other financial entries were recorded. (These will be further examined in Chapter 5.)

Some bankers believe that the above described "on line" (always electronically interconnected), "real time" (with response sufficiently fast to affect the transaction itself) operations will not be economically practical during peak load periods such as at the height of the Christmas shopping season. They predict that the smaller transfers at those times will be handled overnight by "batch processing" (in groups of data accumulated and stored for processing until the lines and equipment are clear). But even so, the substance of the activity remains the same.

The advantages of this kind of checkless/cashless transaction are tremendous. The bank-check problem is solved. The delay in paying checks is eliminated. ("Playing the float" is a fine old American custom, and this may require some who have counted on the delay over weekends and holidays to change their ways!) The vendor's need for working capital is reduced because of this prompt payment, and the savings can result in lower prices, higher profits, or both. The barriers to new companies entering the marketplace can be reduced to a similar extent. Interest charges begin at the moment of sale, so those who want or need credit will pay for the service, and those who pay cash can benefit by their promptness. In a store with a lower-priced but larger inventory, a hand-held reader or "data wand" will read the data coded on magnetically stripped merchandise tickets, credit cards, and employee identification badges. With one pass of the wand across the ticket, price and merchandise information will be entered automatically into a computer terminal, updating sales records and producing an itemized sales slip. This will reduce both error and labor needs even further. The flow of funds will be accelerated and the costs of marketing distributed more equitably among those who receive the products. And, not to be overlooked, the elimination of cash will help control our street-crime problem.

But—the ever present "but"—the checkless/cashless society also has its problems. To this point in our lives, credit cards are only a luxury and not essential to the good life. Within very broad limits, their dissemination is a private business matter. Master Charge or BankAmericard condition their issuance of cards on whatever disclosures they consider appropriate, issue cards to whom they choose, and refuse cards to whom they please. What happens when the bank card becomes essential to existence, or at least to the good life, in our computerized society? How does the impoverished, the deviate, or the just plain freedom-loving man who wants information about himself kept private, buy gasoline at night in White Plains (at least if he doesn't have the exact change), or perhaps even groceries during weekday daylight hours?

I recall traveling to the Hilton Hotel in Washington with a good friend and colleague from Michigan who simply doesn't believe in credit cards. He was in charge of the project on which we were all working and had made the reservations for four of us who were together. The rooms were all ready. We other three registered and got our rooms; he could not. The desk clerk had instructions not to accept cash, check, or anything except an approved credit card (apparently because of fear that telephone, room service, or other additional charges wouldn't be paid, even if the room charge was paid in advance). It may have been embarrassing to my colleague that one of his juniors had to vouch for his credit. Yet otherwise he would not have been given a room.

Much the same thing happened to America's veteran silver-haired diplomat, David K. E. Bruce, now U.S. envoy to Peking. He also carries no credit cards. Upon his return from the Vietnam peace talks in Paris, he sought to rent a car from Hertz at Washington's National Airport. The clerk refused even to accept a $400 cash deposit and was unimpressed by Bruce's State Department credentials. The policy was laundry-ticket clear and absolute—no card, no car.

At least my colleague and Bruce had consciously chosen not to have a credit card, and some might contend that their problems were their own doing. There are many persons, however, who simply cannot get credit cards, either because they have no credit or just don't fit the credit-card issuer's profile of the good credit risk. Certainly the poor and the female qualify for this excluded category. At one time or another, so too would the Communist, the black, the Jew, the homosexual, and just about anyone else who isn't traveling down the middle of the road or considered to be "normal." As my 25-year-old nephew, who can't get any credit card despite a quite substantial inheritance, said, "Everyone's pressuring me to get married and stay in the same job for a whole year." I may not agree with his carefree way of life, but he is old enough to have a right to make his own decisions.

There are many who for other reasons do not want to furnish the information about themselves which the credit-card issuer requires. These are not just nonconformists, although even nonconformists deserve admission to the checkless/cashless society. Many of those traveling on the periphery of society, or trying to change it, find themselves constantly under attack from the Establishment. The Black Panthers, the early leaders of the antiwar movement, and the proponents of the drug culture all qualify in this category. Can anyone imagine them making (or even being able to make) the truthful disclosures which the present credit-card application requires? And can one imagine the pleasure of law enforcement personnel at uncovering a false statement, which may be a felony?

An uncontrolled checkless/cashless society threatens to mechanize humans just as it does cash and checks. Unless rights of entry are built into it, the need for the card of admission can be so great as to persuade by far the greatest number of applicants to conform to the norm—which is the credit card issuer's idea of the good credit risk. For the first time in history, we have enough material wealth to create real leisure and opportunity for dramatic, order-of-magnitude intellectual and emotional advances for everyone. Yet we may be fabricating that material wealth so as to deny us the differences of view and attitude which are essential to such advances.

Control of the checkless/cashless society need not necessarily mean administrative regulation by government. A large national credit-card issuer is certainly a form of computer utility. It would not be straining the computer utility bill of rights to add a fourth rule of law:

Computer Utility Rule 4: A computer utility credit card shall not unreasonably be withheld from any individual.

This would create a new substantive right against economic discrimination, similar to the rights referred to in Chapter 2 and to those which are protected by the present laws against discrimination by reason of race, religion, sex, or age. Just how it would be applied would be determined on the same case-by-case basis as these other laws. But for example, an individual (such as my nephew) posting adequate assurances or security with a credit-card issuer might by this rule become entitled to a credit card with a related credit limit, even if he uses a fictitious name and discloses nothing, or changes cards every twelfth day, or does whatever other peculiar thing he chooses; or the information furnished by nonconformists under protest might become protected against disclosure for other than certain very limited and socially approved purposes.

This chapter has dealt only with the access aspects of the checkless/cashless society. The accumulation of huge banks of data about people is also necessarily involved—in theory at least, the individual's EFTS trail makes possible instant surveillance of his every financial move. Already one local police force has a cooperative arrangement with a credit-checking agency; the police exchange current criminal charge, arrest, and conviction records for information as to the locations in which wanted persons are seeking to make credit purchases, and, sometimes, for help in delaying such persons until an arrest can be made. This capability poses a further and different threat to freedom, requiring other protective measures, which we will discuss in the next chapter. Yet if we can design this electronic marvel of the checkless/cashless society, surely we can build into it the limitations, data restrictions, and use controls which will ensure that it does not interfere with other values we treasure. The first and prime requirement is that we *think* about what we are creating as we build it, and not continue to rush unknowingly into the abyss simply because the lure is so great.

4 THE DATA BANK— CIVIL RIGHTS AND CIVIL WRONGS

At 12:40 P.M. on Thursday, November 14, 1957, Sergeant Edgar Croswell and Trooper Vincent Vasisko of the New York State Police, and Agents Arthur Rustin and Kenneth Brown of the U.S. Treasury Department's Alcohol and Tobacco Tax Division, drove up to the home of Joseph Barbara in the little upstate New York hamlet of Apalachin, New York. They were conducting a routine surveillance based on what they considered recent questionable activity and long-held doubts about Barbara, resulting from his criminal record, his association with known criminals, and their police "sixth sense."

What they spotted immediately aroused their suspicion that another gangland conference similar to that held in Cleveland in 1928 was taking place. Expensive Cadillacs and Lincolns were parked in the driveway. They saw a great number of persons, all men, with the dress and appearance of city dwellers rather than rural residents. When the attendees spotted Croswell and the others, there was considerable activity. Several well-dressed men began to run off across the fields and through the trees, away from the police. Others went inside, and many ran to their cars and drove off.

The troopers and agents promptly set up a roadblock on the road leading from Barbara's home, and stopped each of the cars trying to leave. They also radioed their suspicions to the nearby

Vestal police barracks, and help was sent to stop and identify and question those who left the property on foot. A total of 63 attendees were positively identified either at the roadblock or in nearby areas. Few volunteered any information. Vito Genovese, a widely known organized-crime leader later convicted of narcotics violation and now deceased, told Croswell, "I don't have to talk to you, Sergeant." Those who spoke asserted that no meeting or affair had been planned, and there was little apparent pattern to the information received during interrogation. The subjects told a wide range of stories such as Joseph Profaci's "I got lost with my brother-in-law, Joe Magliocco, on my way to a business appointment in Wilkes-Barre" (the same kind of excuse he had given when queried after the Cleveland meeting in 1928). Buffalo City Councilman John Montana, who the year before had been named *Man of the Year* by the Buffalo Police Department, claimed that he stopped in for emergency repairs because his car had broken down while he was driving from Buffalo to New York City. No one was held, and no apparent crime uncovered.

By the next day, law enforcement agencies throughout the country began to appreciate that Croswell and his colleagues had in fact exposed a top-level gangland conference of impressive significance. The attendees came from fifteen states including California, Texas, and Florida, and from Cuba, as well; they had massive criminal records (convictions as well as arrests), and were all either reputed syndicate leaders in their own areas or henchmen and hangers-on. Immediately police, federal agents, sheriffs, state troopers, district attorneys, United States attorneys, state licensing commissions, federal and state legislative committees, and a host of others including newspapers and civil committees, began to inquire into what had happened. These separate groups assembled an enormous number of documents such as hotel registers, air travel reservations, automobile rental slips, telephone charge records, canceled bank checks and deposit slips, and food and drink purchase receipts, all of which reflected planning for the affair, routes taken, preliminary get-togethers of smaller groups of attendees at Syracuse, Wilkes-Barre and elsewhere—just about anything which would show what had and what had not happened. They also began to question the attendees formally and informally, and the resulting statements, affidavits, and transcripts covered thousands of pages.

None of this scattered, separate information seemed at first to prove very much beyond the confirmation that those who had said anything on November 14 had not told the truth.

On March 1, 1958, I was appointed special federal prosecutor under United States Attorney General William P. Rogers, with the assignment of setting up a kind of short-term "little Hoover Commission" (Herbert, not J. Edgar) to find out what was wrong with syndicated crime enforcement in the United States, and recommend improvement. The work of our Special Group on Organized Crime in the United States concluded with recommendations which I still believe would result in major additional advances in syndicated crime enforcement if fully adopted. But the only part significant to this book is our assembly and analysis (without computer) of this unprecedented volume of documentary material located all around the country and even abroad.

Gradually, as the little bits and pieces of data came together, we saw the clear outlines of conspiracy to obstruct justice and commit perjury (and, incidental to the story here but of key significance to the main prosecutive assignment then, concluded that it was the splintered character of American law enforcement that made prosecution of syndicated crime so difficult). The method of proof of the conspiracy was simple.

First, we had the many statements made by the Apalachin attendees to Croswell, Vasisko, and the others on November 14, 1957, each of which could be proved to be false by positive evidence ranging from the testimony of the Barbara maid, who swore she heard John Montana report on arrival, "I'm sorry I was late," contradicting his statement that he hadn't intended to come at all, to retail store records showing quantity purchases for a meeting alleged to be spontaneous and unplanned, to hotel records showing that those who claimed to have been lost had in fact stayed with others the night before and had driven up together. There was a pattern of consistency to the falsity ("I was visiting a sick friend"), however, which evidenced concert of action.

Second, we had proof of subsequent meetings of groups at hotels and motels after November 14 and just preceding grand jury and other appearances. Thereafter the *members* of these groups told stories which differed from what they had said earlier, and which were now more consistent with one another, but were again demonstrably false.

Third, we had proof that the inconsistencies in stories between *groups* arising out of their November 14 statements were eliminated in the subsequent testimony, and that the new and now also more consistent group stories were equally false.

In other words, we proved a series of inconsistent lies, followed by meetings, followed by consistent but new and different lies. The evidence was sufficient to convince a court and jury, and obtain convictions and close-to-maximum sentences. The value of the methodology was established, despite the subsequent reversal of the case because the appellate court concluded more evidence was required to convict. (Contrary to some lay misimpressions, the reversal was *not* because of any infringement of the right to assemble or to remain silent. Great care was taken throughout the prosecution to avoid any such problems.)

The Apalachin prosecutions took place before computer techniques were available. With computers, the year-long effort of 20 attorneys and many supporting investigators could well have been completed in far less time with greater efficiency and control, so that the reversal on appeal might have been avoided. Law enforcement is rapidly coming to appreciate this tremendous potential of the computer to deal with and help solve the increasingly complex problems of organized interstate crime as well as of large-scale white-collar fraud, conspiracy, and economic crime. The Senate Watergate Committee's apparently successful use of computer retrieval techniques is recent indication that the computer clearly will be a major boon to the criminal prosecutive effort.

The thalidomide tragedy furnishes another example of the benefits the data bank can provide. From 1958 until late 1961 this German-developed drug was marketed widely as a tranquilizer, primarily outside the United States. Among drugs that induce sleep or relaxation, thalidomide initially seemed a breakthrough, largely because it was "suicide-proof." One man was reported to have swallowed 140 pills, slept for several days, and awakened with merely a bad hangover. The science of teratology was then in its infancy, so that no one appreciated that the drug was a teratogen and caused birth defects of the most horrible character. When discovery came, the dimensions of the tragedy were almost beyond belief. In Britain alone there are about 100 children who are completely limbless, have hearing and sight defects, and suffer from other injuries as well; about 160 children are armless or legless, and also have other handicaps; about 40 are deaf, without any ears, or are suffering from facial paralysis; the balance of the 433 victims have a variety of lesser deformities, including odd-shaped or missing fingers and toes. A good part of all this might have been avoided with a computerized reporting system keying birth defects observed to all drugs taken by the mother.

These examples demonstrate the great value which may attach to the combination and analysis of large numbers of facts which seem irrelevant standing alone, such as the record of a long distance telephone call between two numbers at a certain time in combination with a name on a remote hotel register, or the identity of a drug taken by a pregnant woman later bearing a deformed infant in juxtaposition to a similar episode at a distant hospital. The conclusion from just one or two of these observations may seem obvious enough in retrospect, but the two years of inaction preceding the Apalachin prosecution—and the tragedies before the thalidomide discovery—show that this is Monday-morning quarterbacking. The staggering number of facts in our daily lives is simply impossible to analyze fast enough with human brainpower alone, and we just don't put these things together mentally until the quantity becomes numerically overpowering. The tremendous memory capacity and amazing speed of the computer, however, now make it possible to retain as much information as can practically be desired and to manipulate it for a wide variety of purposes. Law enforcement is already able to pinpoint potential crime areas so as to predict time and place of a future violation with a high degree of success. A United Nations medical unit has proposed a data bank of birth records, so that any unusual pattern of birth defects can be spotted promptly and related to some specific cause such as the use of a new drug or unusual radiation hazard. Although other applications may be years away, similar banks of information may one day make it possible for a psychiatrist in later years to study a patient's childhood training and habits and perhaps find clues to treatment, or for scientific researchers to make long-term studies of the diets, habits, and other characteristics of persons with specific diseases or health conditions and to find causation and cure.

Lest it incorrectly appear that the computer can do anything and everything, caution suggests that it be added that no presently anticipated computer is cost-effective enough to make it practical to check *all* possible paths in even a moderately large program. It would therefore be a mistake to expect a computer to find every conceivable relationship among the data items in a mass data bank. Integrated data banks (such as a bank of medical or crime information) provide unique opportunities for analysts to find what they are looking for; purely random correlations are not likely to be sought. Despite these limitations, however, the possibilities the computer offers for human benefit boggle the mind.

CHILLING EFFECT OF DATA BANK

But, concurrent with these probable advantages, the accumulation of such huge dossiers of information has a most chilling effect on freedom and creates an enormous potential for misuse. Indeed, the damage resulting from potential *misuse* of facts is far less serious than that which stems from the individual's simple knowledge that the data bank exists, an awareness which has a subtle but inevitable effect on how he conducts himself. Once "misuse" has been defined (a substantial task which has not yet been adequately initiated) it should be controllable in much the same fashion as other forms of antisocial conduct. In fact, enforcement should be simpler because controls can be built right into the computer itself—controls such as the deletion of outdated records and limitation of access to authorized users. Data-bank misuse might therefore be more easily dealt with than other crimes. But the individual's knowledge that every telephone call

made, every airline ticket bought and every bank check written, is (or might be) available for whatever purpose ten, twenty, or more years hence, will subtly affect his choice of friends, the places he travels, and just about everything else he does. The chances are good that most people won't even realize that they have modified their conduct to conform to the accepted standard or norm, however much they do so.

THE GOVERNMENT AND THE GOVERNED

The greatest single danger posed by the mass data bank is undoubtedly the power it vests in the central government. "Knowledge is power," said Francis Bacon, and Lord Acton added that "power tends to corrupt and absolute power corrupts absolutely." The computer's ability to approach absolute knowledge can thus make possible an approach to absolute corruption.

The Watergate crimes furnish powerful evidence of the corrupting influence of the desire for knowledge. The misconduct, reaching into the highest public office, stemmed either from a desire for knowledge or the desire to prevent others from gaining knowledge—the original break-in at the Democratic National Committee headquarters was alleged to have been aimed at collecting information about the rival party; the White House "plumbers' group" was set up to stop information leaks. It is ironic that much of the misconduct resulted from a desire for knowledge of relatively minor significance by any test, for it related to the search for information related to a forthcoming election which most persons regarded as pretty much "in the bag" for President Nixon.

Yet former Attorney General John N. Mitchell still justified his conduct by the need for knowledge. He testified, ". . . I still believe that the most important thing to this country was the reelection of Richard Nixon. And I was not about to countenance *anything* that would stand in the way of that reelection." When asked, "Anything at all?" he is reported to have replied that if it had come to "treason and other high crimes and misdemeanors"— cited in the Constitution as grounds for impeachment of a President—he would have perceived a "very definite breaking point" in his fidelity to the candidacy of his friend and former law partner, President Nixon.

Mitchell also testified in another connection "that it might even have been better, Senator, as you say, take them [the Watergate miscreants] out on the White House lawn; it would have been simpler to have shot them all and that would have been less of a problem than has developed in the meantime." Had the information sought been really essential to the reelection, one wonders whether this feeble attempt at gallows humor might not have been comedy at all but rather reality and even greater tragedy.

Lest anyone conclude that such misconduct is confined to one end of the political spectrum or to people who are "bad" in some strict moral or ethical sense, it should be added that the Vietnam protests which preceded and contributed to the climate that produced Watergate, were also characterized by criminal conduct in which even ordained ministers of religion participated. No, we cannot consider ourselves immune from official or other misuse of a mass data bank just because a law prohibits it—too many people at all levels and walks of life, from the ministry to the highest public offices, consider that "the end justifies the means." Recent Arab terrorism shows once again that those motivated by ideological goals can be much more of a threat than those seeking financial gain.

The problems presented by the data bank, by no means limited to the much-discussed invasion of privacy, were with us long before the computer. However, the computer so compounds the danger as to create issues of a different kind as well as of a more serious degree. How to make it possible to enjoy the tremendous benefits of data collection without endangering personal freedom and privacy in the computer era, presents one of our thorniest social problems. Fortunately a number of computer professionals, legislators, educators, social scientists, and others are aware of and concerned with this computer problem. The National Academy of Sciences' three-year study and 1972 report entitled *Databanks in a Free Society*, led by Columbia Professor Alan F. Westin, is an outstanding analytical effort, and Harvard Professor Arthur R. Miller's 1971 book, *The Assault on Privacy*, helped greatly to focus interest on the problem. But until now the public and the vast majority of those who ought to be concerned have not been adequately alerted to the issue. Ultimately it must be the public which decides how much freedom it is willing to give up to achieve the benefits of the data bank.

INADEQUACY OF PRESENT DATA-BANK PROPOSALS

Until the public is brought into the debate and is sufficiently concerned to authorize new, imaginative and completely different approaches, those dealing with the data-bank problem will be forced to continue to cope with it along the restrictive lines permitted by existing law and precedent. These center around such concepts as libel, slander, unfair competition, property-ownership rights, privacy as embodied in Fifth Amendment privilege, search and seizure, and similar restrictions. Helpful as these tools may be, they ignore key twentieth-century areas of current and future societal impact.

Too often the analysis of data-bank issues ignores the crucial central question of the threat to freedom created by the very existence of the information. Instead it focuses concern:

> on the control of information which is inaccurate or incomplete, to the exclusion of data which is accurate; or

> on the control of data concerning individuals, rather than data-in-gross; or

> on the control of data obtained surreptitiously, as distinguished from that obtained as the result of voluntary disclosure; or

> on the control of data which is subjective and of an opinion character, rather than data which is objective and factual; or

> on the control of data in the files of government data banks in contrast to that in private hands.

Useful as each of these treatments may be, none is of more than limited assistance in penetrating to the core of the problem, which is the simple existence of the data bank. Neither are all of these considerations taken together adequate to the task. Let us consider each approach.

ACCURATE VS. INACCURATE DATA

To the limited extent that the public has evidenced real databank concern, it has been to obtain assurances that the files contain information which is accurate and complete. This is the significant aspect of the Federal Fair Credit Reporting Act of 1970. Although that statute contains some restrictions on the collection and use of credit data, its main value is in the right it gives the individual to know the substance of the information about him in a credit file and correct it, or at least to require the credit agency to include his side of the dispute in future reports. Accuracy is also the thrust of most other legislative and administrative proposals. Even the protection of such traditionally confidential information as medical data has taken a back seat.

Absolute accuracy and completeness (impossibilities in any event) are certainly relevant concerns. But they are not the center of the *computerized* data bank problem; they have been around a very long time, and are the subject of such common law torts as libel, slander, and unfair competition. It may be entirely accurate but still considered socially undesirable, for example, to report that a job or credit or insurance applicant

- has withdrawn only sex manuals from his library;
- once received shock treatment for mental disorder (this disclosure regarding the Democratic Party's 1972 Vice Presidential candidate, Missouri Senator Thomas F. Eagleton, ended his candidacy in record time);
- always votes as a jury minority of one to acquit black defendants;
- has pleaded the Fifth Amendment;
- was arrested (but acquitted) for breaking and entering while still a child;

- resides and has always voted in an election district which unanimously votes Republican or Democrat (when applying for a job with an administration of the opposing party).

The personal emotions I described in Chapter 1 were not because of concern over erroneous data—my father *did* sign the Stockholm Peace Pledge in 1945 and I *was* reading *Life* magazine's critique of Senator McCarthy during my plane flight to Washington in 1953. My feelings were because such trivia might later be used in a remote and unrelated context to evidence something (valid or invalid) which I considered my own strictly personal affair.

It is not the accumulation and reporting of accurate but "suspect" facts which are the key problem—it is the mere fact that the data exist and may be available. An individual may not measure his conduct and limit his action out of concern that it may be *mis*reported, but because it may be reported, preserved and later spewed forth in combination and association with a mass of other facts about himself which reveal his inner soul.

In early February, 1973, there was a meeting on the privacy issue at the National Bureau of Standards in Gaithersburg, Maryland. Some twenty persons were present, all eminent and respected leaders in their fields and all deeply concerned with the need to take appropriate prompt remedial action in this area. A tape recorder in the corner quietly and unobtrusively recorded the discussion. It had been agreed that this would be valuable for possible future research and analysis. Yet when one of the participants came to report, he began by requesting that the recorder be turned off. His report was clearly the most detailed and specific of all, so his request required the taking of copious and necessarily less-than-complete notes by those interested; nothing he said was improper; he did not express concern that the tape would be edited, misused, or fall into the wrong hands, and this was not a realistic possibility in any event. No, these were not his reasons for the request. He simply would not have felt free to speak as he did, with the knowledge that every word and tone was being preserved. Each of us might similarly measure his conduct with care if he were aware that every action were "on stage."

The names of books borrowed from libraries might change radically if every withdrawal was preserved and collected in one central file available for review in connection with future job applications, credit requests, election candidacies, or other endeavors. Recrimination may also be a contributing factor, but one important reason the secrecy of the election ballot has been preserved all these years is precisely because our society understands that a man will not feel free in the election booth if his employer or even his wife and friends can check on who he votes for.

The heart of the computerized data bank problem, then, is the chilling effect on freedom of the mere preservation of large volumes of data, even where accurate, and even if controls on future use are imposed. For this reason any ultimate solution must contemplate restricting mass data-bank accumulations except and only to the extent that identifiable anticipated benefits clearly outweigh the inevitable harm.

DATA-IN-GROSS VS. TRANSACTIONAL DATA

Many analysts of the data bank problem carefully distinguish research and statistical data ("data-in-gross") from that which is identifiable by individual ("transactional data"). Opinion surveys and polls, trade association statistics, and motorist driving surveys obtained without identification of the respondent are examples of data-in-gross; credit, employee personnel and FBI crime-investigative files are transactional files. Many files, such as those of the census bureau, are hybrids, because they accumulate and maintain information by individual but also process and report data-in-gross.

Those who propose data-bank solutions often concentrate on individual files, and either ignore or suggest only the barest minimum of restrictions on data-in-gross. The July 31, 1973, report of the Department of Health, Education, and Welfare's (HEW) 24-person Advisory Committee on Automated Personal Data Systems clearly recognizes the difference between these kinds of information. However, perhaps because of the nature of the Committee's assignment, even here the highly specific and detailed procedures proposed in its excellent report focus on protecting information identifiable by individual. The concern regarding research and statistical data as such is largely only that it be available for competitive analysis.

Again the distinction fails. Data-in-gross can impact the individual almost as much as his own personal or credit file. Indeed, its stigmatizing and stereotyping effect seems more distasteful and unfair in some respects than transactional data, because the individual has so little control over its existence. Thus a person with a long criminal record has ordinarily created or been responsible for the history which leads to his becoming a suspect, even though there may be valid objections to use of the record as discussed below. But the individual who falls into the classification "black, male, 19, resident of Bedford-Stuyvesant, unmarried, unemployed, grade school education only, father unknown" becomes a suspect only because that is what the statistics tell us.

Few people seem to realize how enormously we are already affected by data-in-gross. Our fire insurance and automobile premiums are fixed by unrelated experience in the areas in which we reside, in large part even to the exclusion of individual experience (State Farm Mutual Automobile Insurance Company charges the short-distance commuter about a 10-percent lower rate than the long-distance commuter who lives 10 miles or more from where he works, even if the latter drives to work only once a week and uses his car far less than the former); life insurance premiums are calculated by age and sex, largely irrespective of individual health and condition; bank credit, interest rates, and availability of credit cards are set by age, employment, length of residence at present address, and the like, frequently apart from individual integrity and ability. Trial lawyers develop an acute sense of "smell" as to how potential jurors with certain characteristics will vote, and they use this training to pick favorable juries; a *New York Times* article reports that auto insurers often refuse to sell collision insurance to divorced women, not because they have higher accident rates—they don't—but because their lawyers believe that juries will vote against divorcees in accident cases. Social scientists have achieved a high degree of accuracy in predicting future patterns of criminal conduct among children and recidivism among released prisoners predicated primarily on background information, to the point where one's belief in individual self-determination is shaken.

Unlike transactional data where individual rights have long been the subject of some common law concern, there is almost no legal precedent to help us handle data-in-gross. Nothing in present law, for example, prevents the politician from statistical research and analysis of voting patterns and instruction to his workers *not* to "get out the votes" of certain specific categories of person because a high percentage always votes the other way, nor prevents the stigmatization of a race or religion on the basis of patterns of mass behavior. The problem is that the computer makes so much more of this possible, that data-in-gross can ultimately become as much of a concern as individual data. I do not mean to suggest, of course, that statistical analysis is necessarily dangerous or that the collection of data-in-gross should be forbidden—far from it. But concentration of controls on the transactional data bank alone necessarily ignores a crucial part of the privacy problem.

VOLUNTARY VS. SURREPTITIOUSLY OBTAINED DATA

There is a third approach to the data-bank issue. That is to restrict the collection and use of data obtained surreptitiously, imposing lesser control on that furnished voluntarily by the individual. Again the distinction is of limited value.

Data must be obtained surreptitiously in certain kinds of cases where the integrity of the subject is at issue, such as criminal investigations, some kinds of personal and credit checks, and the like, or where knowledge of the observation will itself affect the value of the data to be collected, such as in the investigation of mentally disturbed patients or children at play.

Moreover, society is coming to realize that the distinction between what is voluntary and what is not may be more apparent than real, because so many persons are the unavoidable products of their environments. What may seem "voluntary" to one person may well have been inevitable to another. This was brought home vividly during the discussion of this subject by the students in my Columbia Law School seminar on "Computers, Society and Law." A white student, with an obviously relatively secure and trouble-free background, argued vigorously in favor of the right to preserve voluntarily produced data, even of an arrest record in early childhood. To him it was clearly lawful, proper, and fair that the individual be faced with what he had chosen to do. A black student listened quietly for a time but then jumped in. He objected vigorously to the suggestion that what had happened to him as a black child in the ghetto was "voluntary" in any equitable sense of the word. He also protested the suggestion that nothing he might accomplish in later years could ever free him from the taint of his background. After listening for only a few moments, most of us acknowledged how different each of us was from the child from whom he had grown, and we agreed that the voluntary-surreptitious dichotomy was not the complete answer.

The above is not to suggest, of course, that there is no value to the voluntary-surreptitious distinction. Certainly data furnished in confidence or for a specific and limited purpose is entitled to special consideration, even where not technically privileged or legally protected from further disclosure (as in legal or medical privilege).

Similarly, data should be accorded special treatment if it is obtained in violation of the individual's anticipation that it is not being observed or recorded (the generally adverse public reaction to the use of secret recordings ordered by President Nixon furnishes telling evidence of the reasons for this) or if it relates to his conduct under earlier or different circumstances. But to suggest that this is the whole answer is again to fail to treat the major part of the problem.

OBJECTIVE VS. SUBJECTIVE DATA

At a recent university seminar, we were again considering this question of the data bank and the difference between the kinds of facts which should be preserved. Two university professors from very different disciplines agreed that one should not seek to distinguish on the basis of some "free choice" quality relating to the facts. They argued instead that the more serious concern should be with regard to the preservation of *subjective* facts—opinion—rather than *objective* facts. If subjective facts were kept out of the data bank, they contended, the privacy problem would be resolved.

But is there really a difference between a subjective and an objective fact? A fact is subjective or objective depending on how it is used. These professors were apparently concerned with the process by which university advancement operates (something which troubled me not at all at the time, so perhaps my own bias influenced my response). But let's examine student reports on professors. A student's report that a certain professor is dull constitutes an expression of the student's opinion—a subjective fact. If the same student, however, is shown to have reported that each and every professor he has had during his four years of college is dull, in contrast to the reports of his classmates, that says something about the student himself, not the professor—it is an objective fact not requiring an examination of the student's mental processes and conclusions. Suppose at 3:12 P.M. on Friday, May 11, 1973, all 1200 former students of Professor Jones mail letters from points across the entire country to the head of his department. All arrive just in time for the moment of debate on his tenure. All state in identical form that he is dull. This may reflect their combined opinions, but it also suggests an agreement or understanding or plan or conspiracy among the students to defeat appointment. The students' opinions are thus at the same time both subjective facts, or comments on Professor Jones's competence, and objective facts, or evidence of conspiracy irrespective of what any student actually believed. In other words, a *fact* is subjective or objective or something else, depending on how it is used. In this light the distinction between categories of facts doesn't really help much to solve the data-bank problem.

Moreover, facts themselves are tenuous creatures. At one time I was an Assistant United States Attorney in charge of defending all claims against the United States in my District, under the Federal Tort Claims Act. Day in and day out I tried negligence cases, mostly automobile accidents of one kind or another involving postal trucks, army and navy vehicles, and the like. Anyone who has gone through such an experience knows that two eyewitnesses rarely see the event identically. Instead they bring to it their whole lives and upbringings, prejudices, biases, what they ate for breakfast, how they feel, the inadequacies of their eyesight, hearing, smell, and touch, their attention spans—in short, everything. The best one can hope to get from several eyewitnesses to the same accident is a kind of general consensus as to what happened. This is usually more than sufficient to resolve the issue at hand but is also convincing proof that a fact, like beauty, is in the mind or eye of the beholder. It is not some absolute reality upon which the world must rely.

Even scientific facts and principles are not immutable. Not so long ago the electron, proton, and neutron were the basic and smallest theoretically possible building blocks of all matter; there was no possibility of more than 92 elements, and the principle of the conservation of matter (not energy) was an accepted law of physics. But these "truths" didn't last. It is not mere semantics to conclude that there is no such thing as a fact that mortals can "know" with certainty, but simply a reported observation which may or may not be absolute truth—if there is such a thing—and that the objective or subjective nature of the reported observation depends precisely on how it is used and may vary from use to use.

There are certain kinds of opinion data deserving of special treatment in a data bank, such as critiques of personal conduct or habit patterns, but concentration upon such data alone also results in the abdication of responsibility to deal with the total problem.

PRIVATE VS. PUBLIC DATA BANKS

Another frequently asserted distinction in the discussion of the data-bank problem is that between private and public data banks. Those making this distinction usually focus regulatory control only on the public bank, on the assumption that it alone poses a serious social problem. This may make sense in terms of traditional concepts of governmental power, but it certainly does not strike at the heart of the data-bank problem.

Fortunately no overall national private credit data-bank clearing house yet exists, but the impact of such a private file on the individual would be very much the same as a government-operated one. Moreover, unchecked, its contents would be as subject to government exploration and subpoena as any other private information. If Congress was right to become so concerned about a government central data bank in the late sixties, it should be equally concerned about the potential to accumulate and concentrate the same types of information in private hands. Perhaps Congress should be even more concerned, because it would have less power to monitor the private activity than the activity of a bank operated by a branch of government.

Nor does the data-bank problem break down cleanly along traditional public/private lines. For example, *The New York Times* has already computerized its newspaper morgue, and has made it generally available for a reasonable price. Placed into full commercial operation in May, 1973, the three-million-dollar information bank then contained three and one-half years of *Times* article abstracts, plus a variety of abstracts from 65 other publications. This beginning data base consisted of nearly 400,000 *Times* records covering the period between October, 1969 and March, 1973. Each month the daily and Sunday issues of *The Times* yield about 9000 additional discrete items which include virtually all the news and editorial matter and exclude only those items of no discernible research value. Even significant advertisements are entered into the data base.

New entries into the bank, according to *The Times* plan, are to be made no later than 96 hours after publication. In addition, *The Times* expects to gradually work backward to the beginning of this century (the age of its clipping morgue), a job it estimates will take eight to twelve years.

The articles of research value from other publications also being entered into the data base include American and British newspapers, major newsweeklies, business and trade publications, sports magazines, and journals of news and commentary on social, political, economic, and cultural subjects. Items are selected if they (1) deal with subjects not covered in *The Times,* (2) provide more detail, (3) treat subjects from a different point of view, or (4) provide subjects of substantial research value not otherwise obtainable. The Times plans to add publications even beyond these initial 65, if justified by future subscriber interest.

Commercial service is offered by The Times under a price structure that depends on timing and volume of usage. If it is as successful a venture as The Times contemplates, the chances are good that ultimately many—perhaps most—newspapers, plus libraries, government agencies, businesses, and others, will come to rely on it rather than on the more expensive and less satisfactory alternative of maintaining their own morgues. The advantages of such a computerized morgue may be substantial, but the dangers of perpetuating error, opinion, or just The Times' own selection of what is worth preserving and what is not are serious and may even outweigh the benefits. History will be what The Times says it is, not the accumulation of a host of independent researchers each voicing his own views and conclusions. It was a reporter's discovery in a newspaper morgue of G. Harrold Carswell's statement a full generation before his U.S. Supreme Court nomination ("I yield to no man . . . in the firm, vigorous belief in the principles of white supremacy, and I shall always be so governed.") that probably led to Carswell's Senate defeat. Decisions to report and preserve certain categories of facts, such as the early Carswell speech, by thousands of newspapers all over the country undoubtedly balance each other in political and social effect; if concentrated in one or a few hands, they could be of dictatorial significance.

The realities of newspaper reporting and publishing—even for The Times, certainly one of our nation's great newspapers—are also such that a story, once published, often becomes incontrovertible fact and history for that newspaper. Woe betide the candidate whose first news story is an unfavorable one—"Doe doesn't honor his obligations." True or not, this comment is likely to appear in subsequent stories and bedevil him over the years. The same is true of favorable comment and even error. This may be of lesser consequence balanced against freedom of the press where only The New York Times alone is concerned. But if The Times' morgue becomes fact and history for the rest of the press of this country, sad days will have befallen us. And it doesn't make any difference that the history *The Times* thus creates is technically a private one, was not improperly obtained in violation of some person's property rights, and does not violate the libel laws.

SOLUTIONS

Clearly no single approach or remedy will provide a solution to the vast social problems posed by the computerized data bank. But a beginning is certainly very much in order now. And the basis of that beginning must be broad and deep public debate and understanding of the issues.

Minimum standards for any data bank of magnitude should certainly include:

> public notice of the existence, extent and nature of data banks;
>
> clear assignment of responsibility for administration and security to designated identified persons;
>
> rights of access in appropriate circumstances;
>
> correction and deletion of outdated or inaccurate materials;
>
> assurances of security to avoid error and misuse; and
>
> maintenance of adequate records of entry, access, use, and deletion.

Some information, such as secret ballot and jury votes, might be excluded almost universally; information which falls within traditional medical, religious, legal, and informer privilege might be classified at a somewhat lower level, and so on down the line from "sensitive" information, such as a Fifth Amendment plea or an arrest record, through information obtained surreptitiously or in derogation of the individual's expectation, to that furnished voluntarily and ordinarily completely public and "insensitive." Considerations of use would also be pertinent. What a man has had for dinner at home each night for a week might be obtained and analyzed to try to pinpoint the source of hepatitis, ptomaine poisoning, or trichinosis, but not to identify his nationality or religion by the foods he customarily eats.

Outlining these considerations is of course only the beginning of the matter. The enforcement problems are enormous. For example, it can be extremely difficult to establish that full disclosure of data-bank contents has been made as required, particularly because new data may have been legitimately added since a report was rendered. As a result, an audit of a data bank will frequently show discrepancies between a report and the contents on the day of the audit. Rules prohibiting use of data in a bank are primarily effective in prohibiting overt use of the data outside the system, because internal use is so difficult to detect. With respect to the required deletion of inaccurate or outdated records, proof that something does not exist is always problematical, and complete erasure poses special problems with some present computer systems.

Still our system of government relies in large part upon trust and the integrity of our citizenry. Without this our tax laws would long ago have proved completely ineffective. Most operators of data banks will undoubtedly do their best to comply with the rules. Difficult as enforcement may be, it is possible and will improve with advances in technology.

The HEW Advisory Committee, chaired by Willis H. Ware, a computer scientist with great vision and enormous concern in this area, proposed in its 1973 report a number of basic principles as safeguard requirements for automated personal data systems. These are:

> There must be no personal-data record-keeping systems whose very existence is secret.
>
> There must be a way for an individual to find out what information about him is in a record and how it is used.
>
> There must be a way for an individual to prevent information about him that was obtained for one purpose from being

used or made available for other purposes without his consent.

There must be a way for an individual to correct or amend a record of identifiable information about him.

Any organization creating, maintaining, using, or disseminating records of identifiable personal data must assure the reliability of the data for their intended use and must take precautions to prevent misuse of the data.

The report takes a firm stand against the establishment of any standard universal identification scheme (the *s*tandard *u*niversal *i*dentifier or SUI) using the social security number, without first imposing stringent safeguards against possible abuses of computer-based record-keeping systems. It recommends in this connection congressional action giving each individual the right to refuse to disclose his or her social security number to any person or organization that is not authorized by a federal statute to collect and use the number, and that organizations with authority to use the number be prohibited from disclosing the number to organizations that lack such authority.

The report also recommends:

> Federal legislation guaranteeing individuals the right to find out what information is being maintained about them in computerized systems, and to obtain a copy of it on demand.

> Legislation authorizing individuals to contest the accuracy, pertinence, and timeliness of any information in a computer-accessible record about him.

> Legislation requiring record-keeping organizations to inform individuals on request of all uses made of information that is being kept about them in computerized files.

The Advisory Committee's recommendations are valid safeguards to freedom, albeit minimal ones. Despite the fact that they deal with only a part of the mass data-bank problem, much of their substance should be adopted promptly. The President's 1974

State of the Union Message states that this area is being given a high priority by the Administration, so that some early action seems likely.

Also late in July, 1973, the Federal Trade Commission proposed a number of changes in the four-year-old Fair Credit Reporting Act (FCRA). These were primarily designed to put more teeth into the provisions seeking to assure the accuracy and completeness of data files. At the present time, for example, subjects aren't allowed to view their files when they visit a credit data bank. Instead, the contents of the files are read to them. Sheldon Feldman, the FTC's assistant director for special statutes, told the House Banking Subcommittee on Consumer Affairs that subjects complain that as the proportion of sensitive information in the file increases, the extent of full disclosure diminishes. The FTC's principal suggestions were:

> Credit subjects should be given the opportunity to personally inspect and copy the contents of their files, and reporting "codes" used by credit data banks should be translated into layman's language for inquiring subjects.
>
> Credit data banks should be required to get a subject's express authorization before preparing a full-scale investigative report of his private affairs. If the bank doesn't receive such an authorization, it would be allowed to compile only a noninvestigative report.
>
> The disclosure requirement should be broadened to require a third-party user of a credit report, who subsequently uses that report to make an "adverse" decision, to furnish a copy of the adverse report to the person involved. Any adverse decision, including such things as a refusal to rent an apartment and a denial of a license, would trigger the disclosure requirements.
>
> The Civil Service Commission and other government agencies performing employment-reporting functions should be included within the definition of a "consumer reporting agency."

Feldman also said that "it has become apparent to us that until the FCRA is strengthened from the standpoints of coverage, clarity, and liability, its enforcement will remain difficult and unsatisfactory, and its goals largely unfulfilled." Among other things, Feldman suggested that Congress beef up the Act's civil liability sections. At present, he maintained, these sections "do not appear to be effective to deter noncompliance. The chances of recovery of damages under the act are sufficiently remote, and the amount of recovery so insignificant, that private legal redress is virtually nonexistent . . . To our knowledge, not one dollar of damages has ever been judicially awarded to a plaintiff in a civil suit brought under the FCRA." Feldman said the $100 minimum liability provision of the federal Truth in Lending Act has promoted a high degree of compliance with that statute, and he urged that a similar requirement (applicable only to actions by individuals) would aid in FCRA enforcement.

The FTC's recommendations, although also good ones which deserve consideration, are of far less significance and scope than the Advisory Committee's. They are included here because they were released at about the same time. This was not just coincidence, but the inevitable result of the more substantial number of proposals being advanced with respect to this aspect of the data bank problem. All stem from the fact that in this area there is at least *some* degree of public awareness and concern. Indeed, although unaccompanied by adequate public understanding, the degree of concern is sufficient to have made securing the right to privacy as politically popular and noncontroversial as apple pie and mother love. The result could be that in the rush to protect privacy, other important interests may suffer.

Legislation dealing with privacy, endorsed by members of both political parties, is pending in the Congress, and passage of something appears possible even in 1974. The President has appointed a special Presidential Committee to study the problem and make recommendations on a high-priority basis. It is chaired by the Vice President and includes the Attorney General, five other Cabinet officers, and four additional high-government officials. The speed with which the revelation (about the appeal of the privacy issue) has come is suggested by the special Presidential radio broadcast to the nation on the subject of privacy alone on February 23, 1974, acclaiming the right to privacy as "the most basic of all individual rights"—apparently more important than food, clothing, medical care, and housing; or freedom of speech, assembly, and religion, and "a fair break in the marketplace!" Remember that it was not long before that members of this same Administration set out to bug confidential discussions at the opposition's Watergate headquarters, to burglarize Daniel Ellsberg's psychiatrist's office for privileged medical records, to encourage special intensive audits of the income tax returns of opponents, and to secretly record all Presidential conversations and telephone calls.

As much as action is needed, what may come of such extremes could itself turn out to be extreme. This is not the way to achieve the delicate balancing of conflicting considerations between the right of privacy and the right to and need for information.

The number of proposals extant in the privacy area, however, is evidence that in the other computer societal areas, leadership, analysis, and discussion would surely bring public concern and solutions—hopefully based also on understanding. The greater length of this data-bank chapter in comparison with the others in this book compels the same conclusion. Despite the inadequacy of the effort regarding privacy to this point, here there is at least *some* experience to test and measure, and there are proposals to evaluate and criticize. It is sad that there is substantially nothing in so many other important computer societal impact areas.

CURRENT LEGAL GUIDELINES

It is far too early to expect to be able to have specific and detailed final answers to the data-bank problem; perhaps we will never achieve precise definitions and rules, because the many public policy considerations involved are so powerful and at times in such conflict. Privacy, for example—only one aspect of the data-bank issue—is very much the other side of the coin from the right to information; in many situations privacy and freedom of information are in fact mutually exclusive alternatives. Any effort to eliminate the data bank's chilling potential effect on freedom may come into direct conflict with the police power and the public's right to health and safety. In a clash unrelated to the computer's effect on our commercial lives, the interests reflected in the laws against libel and slander met those represented by freedom of the press head on, and a tenuous and imprecise *modus vivendi* was worked out protecting individuals, *if* they were not public figures, and members of the press, *if* they were not motivated by malice. Congress did not pass the Fair Credit Reporting Act, protecting one kind of privacy, until four years *after* it had enacted the Freedom of Information Act, securing the public's right to certain types of information. And the latter Act has focused on another and in some respects equally difficult problem—that of rationalizing the need for access to information with the protection of property rights, such as business trade secrets and financial data. This is not to say that property rights stand higher than personal privacy, but rather that it is difficult to determine where the public interest lies in the scheme of disclosure because of the vast amount of data collected by government regulatory agencies from the business community (especially in the area of product safety and efficacy).

Free press, free speech, the right to assemble, the privilege against self-incrimination, the right to be secure from unreasonable searches and seizures, due process, police power, and a host of other constitutional provisions are all frequently cited by one side or the other in civil-rights cases. How one balances the opposing interests may very well depend on whose ox is being gored.

Although privacy issues have been with us for many years, there still is little law defining the right to privacy in any detail. Despite what some people think, for example, not all searches without warrant are improper, nor all secret recordings of telephone calls illegal. There is even less useful precedent dealing with data bank problems, which are far broader.

Because the conflicting considerations are so strong, resolution of data-bank issues may ultimately come to turn on a judicial evaluation of applicable community standards—something like the vague and ill-defined (and quite unfortunate, in the obscenity area) pornography test recently laid down by the Supreme Court. Evaluation of the relevant community standard was the approach to the privacy issue taken in the *Stark* case in California in 1972. There a special three-judge federal constitutional court struck down the Treasury Department's efforts to require bank reporting of financial transactions of customers under the newly enacted federal Bank Secrecy Act. The court held that bank customers reasonably expect privacy concerning the details of their personal financial affairs, and that society is prepared to recognize such expectations as reasonable. It expressed the crucial judicial test to be "whether there is any reasonable relationship between the end sought to be achieved, i.e., possible assistance to the government in its investigations of citizens, on the one hand, and on the other hand, the peremptory, sweeping, unsafeguarded reporting provisions which it [the Act] authorizes the Secretary to require." The court then applied the test to the facts before it, and concluded that the government's efforts to inquire into customers' financial transactions with banks under the special circumstances presented must yield to the individual's right to privacy.

The U.S. Supreme Court reviewed the *Stark* case on direct appeal, and had a chance to deliver an opinion of major interest and importance which could have shed new light on this difficult problem. Unfortunately, however, on April 1, 1974, it chose instead to avoid the key issues on procedural grounds. Perhaps reflecting the Court's much more conservative approach to some issues, Mr. Justice Rehnquist, writing for the majority in a 6-3 decision, held (1) that the banks themselves could not assert that their constitutional rights were violated or that the reporting requirements were not reasonable, and (2) that individual depositors and the American Civil Liberties Union, who protested that the reports invaded their privacy, did not have legal standing to raise the question. This was because they did not prove they had engaged in any $10,000-or-more deposits or withdrawals, the only size for which domestic reports are required.

Justice Douglas, joined by Justices Brennan and Marshall, dissented vigorously, charging that the statue required banks to "spy upon their customers." He said:

> It would be highly useful to government espionage to have like reports from all our bookstores, all our hardware and retail stores, all our drugstores. These records too might be 'useful' in criminal investigations.
> One's reading habits furnish telltale clues to those who are bent on bending us to one point of view. What one buys at the hardware and retail stores may furnish clues to potential uses of wires, soap powders, and the like used by criminals. A mandatory recording of all telephone conversations would be better than the recording of checks under the Bank Secrecy Act, if Big Brother is to have his way. The records of checks—now available to the investigators—are highly useful. In a sense a person is defined by the checks he writes. By examining them the agents get to know his doctors, lawyers, creditors, political allies, social connections, religious affiliation, educational interests, the papers and magazines he reads and so on *ad infinitum*. These are all tied to one's social security number; and now that we have the data banks, these other items will enrich that storehouse and make it possible for a bureaucrat—by pushing one button—to get in an instant the names of the 190 million Americans who are subversives or potential and likely candidates. . . .
> Since the banking transactions of an individual give a fairly accurate account of his religion, ideology, opinions, and interests, a regulation impounding them and making them automatically available to all federal investigative agencies is a sledge hammer approach to a problem that only a delicate scalpel can manage. Where fundamental personal rights are involved—as is true when as here government gets large access to one's beliefs, ideas, politics, religion, cultural concerns and the like—the Act should be "narrowly drawn" (*Cantwell* v. *Connecticut*, 310 U.S. 296, 307) to meet the precise evil.

Although the significance of the Supreme Court's *Stark* opinion is limited by its procedural aspects, the tone of the majority opinion clearly suggests that the Court considered that the criminal law enforcement benefits of disclosure outweigh the privacy considerations. This may change. Justices Powell and Blackmun concurred with the majority in this case, but Justice Powell's brief concurring opinion indicates that they might vote otherwise in other circumstances. It approved reporting domestic banking transactions of $10,000 or more but not if that figure were lowered to take in large numbers of smaller deposits and withdrawals. Justice Powell said:

> Financial transactions can reveal much about a person's activities, associations, and beliefs. At some point, governmental intrusion upon these areas would implicate legitimate expectations of privacy.

This latter reference to legitimate expectations of privacy is strikingly similar to that of Justice Douglas, who wrote:

> Customers have a constitutionally justifiable expectation of privacy in the documentary details of the financial transactions reflected in their bank accounts. . . .
> We said in *Katz* v. *United States*, 389 U.S. 347, 351–352, "What a person knowingly exposes to the public, even in his own home or offices, is not a subject of Fourth Amendment protection. . . . But what he seeks to preserve as private, even in an area accessible to the public, may be constitutionally protected." As stated in *United States* v. *White*, 401 U.S. 745, 752, the question is "what expectations of privacy" will be protected by the Fourth Amendment "in the absence of a warrant." A search and seizure conducted without a warrant is *per se* unreasonable subject to "jealously and carefully drawn" exceptions, *Jones* v. *United States*, 357 U.S. 493, 499. One's bank accounts are within the expectations of this society in that category. For they mirror not only one's finances but his interests, his debts, his way of life, his family and his civic commitments.

And Justice Marshall, in his dissent, similarly wrote that "the customer of a bank, having written or deposited a check, has a reasonable expectation that his check will be examined for bank purposes only." All seem to endorse the District Court's approach, and suggest that one day we may have a 5-4 decision to that effect, even with the same Supreme Court bench.

I recall a bar association opinion similar to *Stark*, over a decade ago. I was defense trial counsel to a major over-the-counter securities house and one of its chief executives. Both were charged with fraud and misrepresentation in the sale of unregistered stock of Gulf Coast Leaseholds, Inc. The alleged conspiracy involved many codefendants and hundreds of millions of dollars in securities transactions taking place over many years, abroad and in the United States. This was one of the first cases in which the later much-publicized secret Swiss and Leichtenstein bank accounts were used to conceal criminal acts, and the relevant transactions concerned large numbers of named coconspirators and other witnesses all over the world, most of whom were hostile. I knew that the SEC and later the Department of Justice and the United States Attorney's office had conducted extensive investigations, and undoubtedly had a massive volume of grand-jury and deposition transcripts, witness statements, and similar cross-examination materials. One of the important rules of trial practice is to avoid asking questions of a witness, where possible, unless you know and can pin him down to the answer. The documents the government had were tremendously important in this connection. If a witness testified to a new and different story on the stand, these papers would help the prosecutor create doubt about his veracity.

I could not expect to duplicate the government's efforts or obtain any large part of its evidence, because this was long before the substantial liberalization of discovery and disclosure to defendants in criminal cases. I had investigators who would help me interview the most important of these potentially hostile witnesses and report what they said, but I wanted as much additional documentary protection as possible by way of secret recordings of the interviews. With the magnitude of the alleged fraud and the great number of witnesses in the case, it was inevitable that some witnesses would try to change their stories at trial. The alternative, if no recording was available, would be a swearing contest between witness and investigator whenever a witness' testimony was in substantial conflict with his pretrial statement. This is always a most unsatisfactory trial procedure, because it introduces the additional and extraneous issue of the investigator's motive, bias, and credibility. The jury may simply disbelieve the investigator. Moreover, the investigator's testimony might be considered "collateral" and therefore not permitted at all. A secret recording would avoid all this.

I submitted the question to my local bar association's ethics committee. Using an approach very similar to that of the District Court's *Stark* decision a decade later, the committee ruled in substance that a witness being interrogated without notice of a recording had a reasonable right to believe that there was no recording being made and that to make one therefore constituted a "tacit misrepresentation." Its October 5, 1961 opinion concluding that the use of the secret recorder would be improper has never before been published. It reads in full:

> This is in reply to your letter of August 4, 1961 requesting the advice of the Committee on Professional Ethics on the following:
>
> It is anticipated that hostile witnesses interviewed by lawyers and investigators during the course of a pending pretrial investigation may at trial deny having made certain statements or deny even having been interviewed.
>
> It is desired to use a secret recording device, such as the "Minifon" in current use by certain governmental investigative agencies, in order to obtain corroborating evidence of the statements of such hostile witnesses.
>
> The recording device to be used is designed to prevent the person being interviewed from realizing that his statements are being recorded. However, there would be no affirmative misrepresentation by the interviewer with regard to the fact of recording, other than may be inherent in the design itself. Moreover, the recording would not be used unless the witness denies having made the statement attributed to him. Its function would be to serve solely as a shield to protect the client against improper testimony.
>
> 1. Would there be any impropriety in the use of such a secret recording device under these circumstances by

either the attorney conducting the pre-trial investigations or an investigator acting on the attorney's behalf?

2. Would it make any difference if the witness being interviewed is also a lawyer, assuming that the witness is not now acting as counsel to any of the parties in connection with the litigation with respect to which the investigation is being conducted?

Opinion: The Committee is of the opinion that the use of a secret recording device by or on behalf of the attorney, under the circumstances stated in either of your questions, would be unethical. See Opinions Nos. 624 and 683 of this Committee, reproduced in *Opinions on Professional Ethics,* a Cromwell Foundation publication, available at the Association's library. It is outside the province of the Committee to pass on the questions whether the proposed use of such device would violate any law or whether its use by governmental investigative agencies is proper. Regardless of these questions the Committee feels that the use of such a device by or on behalf of an attorney under the circumstances of your inquiry would be inconsistent with the honor and dignity of the profession. *Though no affirmative misrepresentation may be involved, the Committee feels that there is a tacit misrepresentation which an attorney should not countenance.*

This reply is for your information and will not be published by the Committee. [Emphasis added]

I did not use the secret recording device of course—although after listening to some of the Watergate disclosures regarding the use of secret recordings of all kinds, not justified as a part of the criminal defendant's constitutional right to counsel and preparation of his defense, I cannot help wondering why our bar associations are so hesitant to proceed with respect to that far more aggravated conduct by lawyers holding the highest positions of public trust.

The rationale of the lower court *Stark* and bar-committee rulings is that privacy is very much a subjective concern. It depends on what the individual should be reasonably entitled to expect, based on relevant community standards. Does he have a reasonable right to believe that his checks will not be examined, except in certain limited circumstances? Does he have such a right to believe that his conversations are not being recorded? Does he have a right to believe that his telephone is not being tapped, except pursuant to a warrant issued on the basis of reasonable evidence that he is committing certain serious crimes? If his belief is reasonable in those respects, the information will be protected against disclosure. This is by no means a satisfactory test, but at the moment it may be the best we have.

THE DATA-BANK PUBLIC IMPACT STATEMENT

Whether or not we will ever have clear and well-defined answers concerning privacy and the data bank, it is not too early to conclude that mass data banks can be condoned only where the purpose they serve is of greater social value than the associated danger. Here, unlike the computer utility and checkless/cashless society access issues in which the individual can be expected to enforce his own rights, some form of limited licensing and regulation may be necessary as a condition to its existence:

> **The Data Bank:** *A mass data bank shall be permitted to operate only if the benefits associated with its operation outweigh the related risks.*

This requirement would of course apply to the data-bank aspects of the checkless/cashless society, as well as consumer credit and all other large information storehouses.

The application of such a nebulous licensing test would be difficult, but not impossible. During 1971–72 I was trial counsel in a hearing conducted by an AEC Atomic Safety and Licensing Board, seeking to determine whether a proposed nuclear power plant should be licensed. This was the first AEC hearing under the National Environmental Policy Act of 1969 (NEPA), the issue being whether the environmental benefits outweighed the environmental risks. There were no definitive environmental guide-

lines; within broad limits everyone simply introduced whatever evidence he wished. We probed biology, botany, geology, ichthyology, nuclear physics, ornithology, zoology and just about every conceivable scientific discipline. At the conclusion most were convinced that a fair balancing of pros and cons had been conducted.

NEPA requires an environmental impact statement with respect to all major federal actions affecting the environment. It may already apply to mass *federal government* data banks. A similar requirement that there be an impact statement might be applied to certain types of large *private* systems. Such a data-bank impact statement would contain a formal written description of each proposed data bank before its authorization, including its objectives, activities, and controls, with an analysis and presentation of all anticipated benefits and detriments. Thereafter an appropriate licensing authority would make an evaluation and report whether the benefits outweighed the risks. In contested or close cases a hearing would be held in which the public would participate, with rights of cross-examination and argument. At the end of the process everything would be out on the table for all to see and consider. There would be opportunity at least to communicate public concern and seek legislative redress.

PRIVACY COMMISSION

It is premature to do much more in the computer/society field than identify some of the problems and consider their implications. Except where special problems make early action essential, we are simply not yet ready for solutions. There is not enough raw data, let alone research, analysis, discussion, and debate, to permit adequately considered decisions to be made.

The solutions proposed herein give form and content to conceptual problems so as to stimulate discussion and give people something to shoot at—to paraphrase George Washington Hill, I don't care what people say as long as they talk about the issues. But the outlines of these solutions have been sketched only generally, omitting such difficult and troublesome questions as which kinds of computer utility are to be subject to controls. It is obvious that a giant interstate or international computer utility, purporting to furnish credit information on 100 million people, would fall within the regulatory framework; it should be just as obvious that a small local computer utility would be subject only to the most limited restrictions. There is a massive gray area in between, concerning which proposals should not even be suggested until there has been some sound research.

In the data-bank area, however, the conclusion is inescapable that some kind of at least initial and temporary regulatory control is necessary. Despite my reservations, further discussion of procedural matters is therefore essential. Another regulatory agency in an already over-regulated society is most dangerous—the recent Watergate experience teaches once again the wisdom of avoiding excesses of executive and administrative power. Where control is required, to the extent possible it should be vested in a legislative branch responsive to the people and limited by a constitution; when the legislature has acted, interpretation and enforcement should shift to a judicial branch.

I propose the creation of a Privacy Commission with limited administrative powers, incorporating as many of the legislative and judicial ingredients as possible.

The Commission would have a ten-year life, with instructions to design its own demise. Its operation would be subject to continuing review by a Joint Committee of the Congress, with a formal overall review scheduled for its third and sixth anniversaries. It would be composed of nine outstanding citizens appointed by the President and confirmed by the Senate. These members would be selected so as to include representatives of the public, the legislative, executive, and judicial branches of government, civil-liberties groups, computer scientists, the legal profession, and industry. No public or private data bank in excess of a certain size (measured in terms of numbers of files and interstate impact) could be established or be permitted to operate after a certain period without a license issued by the Commission.

An applicant for a license would be required to prepare a comprehensive Data-Bank Impact Statement, in a form similar to the NEPA statement described above. The Statement would be filed with the Commission, published in the Federal Register, and given especially wide public circulation in the specific logical and geographical areas of its impact. There would be a 120-day period following its publication during which comments would be invited from interested persons. The Commission's staff would also circulate the Statement to all relevant federal, state, and local agencies. During the 120-day period the staff would study the Statement and comments, much as the SEC staff studies registration statements. The staff would prepare its own comments on the basis of its analysis and the views submitted to it. There would be an informal exchange period between Commission staff and data-bank license applicant, during which the applicant might accept certain modifications or limitations. Although the informal exchange could be conducted privately at the discretion of the staff and the applicant, the staff's comments and all proposals and counter-proposals would be made readily available to the public.

At the end of the 120-day period, or any additional period agreed upon by the staff and the applicant, the Commission would review the final objections submitted by the staff to the application as amended, as well as all other objections posed by third parties. The Commission would be required to publish formal written findings, conclusions, and recommendations within 60 days, except if it found that questions of fact were involved; in that case it would hold a public hearing at which the contested issues would be explored in depth.

The Commission would have investigatory powers to police the operations of licensed data banks and ensure continued proper operation. This police power would include a kind of ombudsman function responsive to complaints from the public, and the right to suspend or revoke a license and institute civil or criminal legal proceedings for violation of license provisions or law.

The right of the Commission to deny licenses would be extremely limited. Apart from its prosecutorial role, its powers would come down essentially to public "jaw boning" and recommendations to Congress. Except where an application proposed a program which would be illegal under some existing law, the Commission could not permanently withhold a license. Its most commonly used power would be to ensure that all the facts and issues were thoroughly explored and discussed publicly. In unusual cases it would have the additional limited power to delay the issuance of a license for a 180-day period following the transmission of its formal written decision to Congress. During that period the Congress might consider its recommendations for legislative action.

SWEDISH DATA-INSPECTION BOARD

It is not too early for this kind of action. Sweden already has a Data Inspection Board. Some local West German governments have also moved in the direction of data-bank control, suggesting that the German Federal Republic may be the next nation to do the same.

The Swedish Board came into existence July 1, 1973 and will assume its full duties July 1, 1974. In order to make its decisions as representative of broad public opinion as possible, Sweden's Board is composed of nine members drawn from labor, Parliament, the legal profession, government agencies, the computer industry, and general business. It will eventually also have a staff of about 20–25 employees, including computer experts, lawyers, and business graduates.

The Swedish law that created the Data Inspection Board gives that agency broad powers to control and regulate the maintenance of data banks containing any information on individuals. It is probably the first national law to establish such a government agency aimed at protecting individual privacy.

Under the "Data Act," signed into law by the late King Gustaf on May 11, 1973, no one may keep or start a personal register without explicit permission from the Data Inspection Board. Although registers can be established by the King or Parliament (not by local governments) without permission and these are likely to be among the largest, even here the Data Inspection Board has substantial power. It must be consulted in advance and can regulate any aspects of a government register left open by the government (e.g., security measures). It also has very considerable "jaw-boning" influence, because its advice must by law be given serious consideration and will be reported to Parliament, where the appropriate minister may be questioned.

The Board will give permission to start a system containing restricted personal information unless there is reason to anticipate that "undue encroachment on the privacy of individuals will arise." Where there is some question in this regard, the Board will consult with the individuals whose records are to be kept in the system to determine their attitudes toward it before granting permission.

In the absence of "extraordinary reasons," no one other than the relevant government agency can keep records that involve police matters, psychiatric records, and the like. The Act also states, "Permission to start and keep a personal register otherwise containing information about anybody's illness or state of health or information that anybody has received social assistance, treatment for alcoholism or the like . . . may not be granted to a person other than an authority which is by law or statute responsible for keeping such a record" (unless there are "special reasons"). It also proscribes the keeping of any records on persons' religious or political views without good cause, but permits religious and political organizations to keep records on their membership.

Whenever permission is granted to start a data bank, the Board is required to "issue regulations as to the purpose of the register and the personal information that may be included. If there are special reasons the permission may be limited to [a] certain time." The Board is also empowered to issue regulations on the way information is collected, the way it is processed, the technical equipment used, what personal information may be made accessible, the keeping of the information, and the control and security procedures to be followed by the data-bank operator.

If there is any reason to suspect that any information in a register is incorrect, the law states that "the responsible keeper of the register shall, without delay, take the necessary steps to ascertain the correctness of the information and, if needed, to correct it or exclude it from the register." In addition, if that incorrect information has been given to anyone else, the operator of the system has a duty at the request of the affected person to update and correct those records as well as his own. Data-bank operators are also required to keep the records in the system complete and to supplement any incomplete files.

Any individual has a right to request to see his record kept in any data bank, although that right is restricted to a once-a-year perusal of the file. However, the data-bank operator is obligated to provide the record to the individual without charge for his inspection, unless the operator can convince the Data Inspection Board of the need for a fee in some special circumstances.

Data bank operators are subject to criminal action for unauthorized disclosure of information in their systems. In order to supervise adherence to the law, the Data Inspection Board has the power to enter any premises where data banks are kept and operate the computers at those sites itself. In addition, the keeper of the data bank must provide the Data Inspection Board with any documentation or information on its operation that would help the Board carry out its duties.

If the Board finds "the keeping of a personal register has caused undue encroachment on privacy or if there is reason to believe that such encroachment will appear," it may issue new regulations to be followed by the data-bank operator or it may revoke its license to continue operation.

Anyone violating the Act is subject to a fine or imprisonment for one year, and individuals are permitted to take data-bank operators to court if they feel their rights have been violated. In addition, the Act further seeks to protect the security of data banks by making improper access, alteration, or recording a crime ("data trespass"), as follows:

> Any person who, without authorization, effects access to recording for ADP [*a*utomatic *d*ata *p*rocessing] or unduly alters or obliterates such information or includes it in a register will be sentenced for data trespass to pay a fine or to a term of imprisonment not exceeding two years if the perpetration is not punishable by the Penal Code.

The Swedish statute vests great power in the Data Inspection Board. The Board's first Chairman, Dr. Claes-Göran Källner, is a distinguished Swedish jurist and former permanent Undersecretary of State. Speaking in Stockholm at the end of November, 1973, he explained that the statute's approach is consistent with Swedish law, history, and tradition. For example, to guard against possible government abuse of power, Sweden has long had the broadest possible rules requiring disclosure of government information. These rules are given substance by the principle of "remiss" (circulation), which requires that proposals for legislation be submitted to government agencies such as the Data Board before such proposals are finally considered by the Cabinet. Remiss of course triggers application of the open-records law, if it has not already come into play in some other way, and adds an active element to the passive publicity principle.

Dr. Källner is quite aware of the dangers of broad administrative discretion, and hopes to develop as many specific guidelines as possible during the first year of the Board's organization and study before it begins full operation. Thereafter he hopes to narrow the range of discretion to the maximum extent by developing rules as the Board proceeds on a case-by-case method, ultimately evolving a code of procedures in a fashion similar to the manner in which our American judicial system develops rules of conduct.

Moreover, Sweden of course has a different system of checks and balances than we do. Its Board's ruling can be appealed to the Government, which has substantially unrestricted review power, and its Ministers can be called before Parliament; the ombudsman concept adds a further limitation.

I do not think the evidence in this country yet justifies action similar to Sweden's; it remains to be seen whether it will ever be necessary here—I hope not. But neither is the Privacy Commission–NEPA approach a complete solution to the mass data-bank problem, because it furnishes no standards or guidelines on how to decide the close cases. Regulatory agency control cannot be justified for any longer than strictly necessary; rules which require government permission to take action tend to be socially ineffective. Enforcement of general legislation of the kind I have proposed also presents special problems—the size requirement alone offers obvious possibilities for circumvention.

But at least NEPA requires that everything be public and out on the table. Those proposing data banks will have to explain "why" and *think* about appropriate controls and limitations; those opposing them will have a chance to propose restrictions and changes. The decision will probably be comparatively easy in most cases—either the greater value of the information or the excessive harm from its accumulation will be apparent and compelling. And after a number of close cases have been considered and decided, we will begin to produce a body of understanding and law from which more precise tests will emerge. At that point regulatory control can be dropped. Ultimately we will surely develop something like an expanded Fifth Amendment protection to prohibit the antisocial accumulation and use of data-bank information, and strict legal rules protecting against data misuse or abuse. As we do so, it will be essential to keep in mind that the final objective of all these safeguards must be not only to protect individual freedom, but also to secure society's overriding interest in having the greatest variety of opinions and conduct. The data bank cannot be permitted to reduce us to the single norm and cipher of the machine.

5 POINT-OF-SALE MARKETING AND FREE COMPETITION

I was born and brought up at 317 West 89th Street, on New York City's upper west side. We did most of our shopping for groceries and household supplies at a little Gristede's store around the corner on the west side of Broadway, between 89th and 90th Streets. The sales part of the store was roughly 25–35 feet long and 8–10 feet wide, with a counter about 1½ feet wide.

Even during the great depression of the early 1930's, the store's walls from floor to ceiling were tightly stacked with supplies. With almost every order the clerk had to climb up a little ladder which rolled along a railing to reach something high, using a long-handled pole with squeeze fingers at its end, and bend down to the floor for something else. I recall once being in the rear of the store and seeing another supply area behind the customer section, an area about the same size or slightly larger than the front but even more tightly packed with inventory, because room wasn't required for a counter or for customers to move around.

Despite the fact that our local store seemed to have several weeks of supplies on hand, it was frequently out of something or other, and we had to accept a substitute or go without. Today our huge local supermarkets have only a fraction of that inventory on hand (relatively speaking), but rarely are short of a product. Instead of huge stocks occupying large shelf and warehouse space and tying up capital, effective management and in-

69

ventory control keep shipments of supplies moving in regularly, as needed, to assure adequate but not excessive supply. Computers play a small role in this distribution operation in some of the chains, but their impact has only scratched the surface. When *point-of-s*ale (POS) computerized marketing is fully operational, its effect will dwarf the difference between today's supermarket and my childhood Gristede's.

POS installations are technically simple, well within present computer capabilities, and already operating in limited respects at such huge chains as those of Sears Roebuck and May Department Stores. A special kind of cash register at the checkout counter or "point of sale" is tied in to the store's computer facility, whether on-site in the store or elsewhere. The register maintains a record of all the usual details of a transaction, such as sales person, department, and amount of sale. In addition it records the actual product sold. All of this information is then processed so as to produce reports of sales, cash receipts, sales person's commission, profit and loss, inventory levels, and even profitability by product, department, or otherwise, on as frequent a basis as desired.

It is a simple matter electronically to link the inventory level thus calculated with an instruction to the supplier the moment the store's inventory reaches a certain minimum predetermined level. For example, if a purchaser buys one-half dozen size 16 × 34 white shirts of a certain style, and this reduces the store's inventory below the four dozen which the men's shirt buyer considers essential to have on hand while awaiting a new shipment, an automatic electronic instruction goes to the wholesaler to ship another six dozen. The wholesaler in turn may operate in precisely the same fashion, issuing an automatic electronic instruction to his supplier, the manufacturer, the moment the wholesaler's level of supply falls below his predetermined amount. And the manufacturer, finally, may similarly organize his operation so as to issue an automatic order to *his* appropriate manufacturing department the moment his own inventory falls below a certain level. Using computerized manufacturing techniques, it is even possible to manufacture the shirts to order in accordance with instructions, using numerically controlled machines and involving a minimum of human effort.

All of this is obviously extraordinarily efficient. It means that the store will have what the customer wants on hand when the customer wants it, so that the sale is not lost. At the same time, the store's space requirements for inventory and need for capital tied up in supplies is reduced to the very minimum, making possible greater profit, lower price, or both.

Until now, POS marketing has been almost universally acclaimed for its potential efficiency and utility (once present excessive start-up costs are brought under control) and heralded as the marketing technique of the future. It has been treated as only a simple extension of the management techniques which made possible inventory reduction from levels of the days of my youth to current supermarket levels, ignoring basic differences which can have enormous consequences to our traditional competitive way of life.

Although it is true that POS is well within present technical capabilities, it is extraordinarily complex and expensive and extremely difficult to get to function properly, as are most advanced modern computer systems. A long period of time and a very great deal of money are required for a merchandiser to install such a system and eliminate the inevitable bugs. New York City's most unfortunate computerized off-track betting experience is a good example of the kinds of problems experienced with these systems. Millions of dollars of projected New York City income were lost before a system far less complex than POS was installed and operational. There are far too many other such examples. A store is not lightly going to change its system to tie into a new supplier simply because the new supplier offers a product just a little bit better or a few cents cheaper.

Those who have installed computer controls within their businesses understand the inflexibility it introduces to their operations. Change is possible, but once the system is operational, change to accommodate new and different "interfaces" (connections or links) with the external world will be undertaken only *in extremis*. Because of this, POS introduces a form of structural rigidity into the economic system, which is at odds with the flexibility required for effective free competition.

A more subtle example of the competitive changes which the computer is quietly introducing into our economic structure is what the computer-services industry terms "incremental marketing." This phenomenon encourages the expansion of great companies with computer capabilities into new separate and distinct lines of commerce, interfering with and sometimes suppressing previously existing freedom of trade.

By "incremental marketing" is meant the selling of excess computer capability by a vendor whose primary product is something other than computer marketing. For example, an airline manufacturer or insurance company will install sufficient "in-house" or internal physical computer equipment and hire or train sufficient personnel to be able to handle its own peak-load computer requirements. These peak loads may be very large during some of the main business hours of the working day, but computer needs will be quite low on nights or on Sundays and holidays. Frequently, also, requirements are cyclical. Payrolls, for example, come up only once a week, or once or twice a month. It is natural and perhaps even desirable under pure economic theory for these companies to market their excess off-peak capabilities to customers in other industries who do not have adequate in-house computer installations; more and more of the Fortune 500 are already doing so.

The most direct effect of this marketing is on the independent computer-services companies which offer competitive services to industry. They have neither the influence and power which come from success in another line of commerce, nor the ability to spread costs over two separate markets—one internal and captive, the other external and public. Thus they lose out. This first effect has already reached sufficient proportions to have aroused protests, complaints, and some litigation by the independent computer-services industry.

A secondary (and ultimately perhaps far more serious) overall economic effect may be to permit the incremental marketer to work his way whole-hog into the separate line of commerce in competition with, and finally sometimes even overpowering his former customer. Once the incremental marketer has acquired the initial foothold and then learned the intricacies of the customer's business by servicing its most intimate requirements, the possibility of conglomerate and congeneric acquisition or other entry into the business is apparent. The trend towards merger, acquisition, or domination may be even greater if "facilities management" is involved. In that special type of computer-services marketing, the supplier actually takes over and operates the internal computer division of his customer's business. He is therefore even more knowledgeable about, and in control of, the customer's operations.

The computer-services industry is still a relatively small and young one, and this secondary effect has not yet been substantial. Economists and antitrust lawyers may differ on the extent to which they believe it will in fact become a serious concern. But incremental marketing as a problem already exists. The added possibility of this second effect and of perhaps tertiary and other yet-unconsidered adverse consequences combine to demand consideration of the issue now, before it is too late and the economic course has become irreversible.

Incremental marketing encourages use in a second, separate, and distinct line of commerce of market power formerly isolated to a single line of commerce; it tends to blur the distinction between lines of commerce and may unbalance competitive prices in the secondary line. Again, depending on how the power is used, this can be in conflict with the conditions required for free competition.

Many believe that the underlying theme and philosophy of American antitrust law is that competitive activity among many small entrepreneurs is healthy and good for its own sake as well as for the ultimate consumer. Time and again the courts have condemned licensing and franchising arrangements which unreasonably rigidify marketing opportunities and other improper extensions of market power. POS and incremental marketing are two examples of changes the computer is introducing into our competitive structure which are in direct conflict with that antitrust philosophy.

Although no case has yet arisen presenting the issue, it may well be that the Supreme Court will ultimately decide that a POS customer-vendor relationship which results in the customer being linked tightly to the vendor's product is just as unlawful as other exclusive franchising or licensing arrangements which the

Court has already ruled to be unlawful. In the *Schwinn* bicycle case, for example, the Court condemned certain marketing arrangements between Schwinn and its distributors whenever Schwinn had parted with title to its bicycles; the same kinds of restrictions were upheld, however, where the distributors acted as agents, consignees, or manufacturer's representatives. There is a distinction of course, but by parity of reasoning the Court might conclude that where the POS vendor does not hold title to the products marketed, any POS restraint is similarly unreasonable and unlawful.

By analogy to the *Northern Pacific Railway* and *Fortner Enterprises* cases, the Supreme Court may also hold that certain forms of incremental marketing, in which the vendor's primary line and its incremental-marketing activities are inadequately separated, are the equivalent of unlawful tie-in sales, constitute the unlawful use of power in one line of commerce to restrict competition in another, or involve unlawful reciprocity arrangements. A suggestion that the Department of Justice may one day seek such a result was given by Donald I. Baker, then the Antitrust Division's Director of Policy Planning, in a speech in mid-1973. Although dealing with a different problem, bank mergers, he expressed for the first time the official view that "tying effect" without coercion, the real vice of incremental marketing, is just as serious an adverse economic effect as the coercive tying which has been illegal for many years. Baker said:

> Bank affiliation with a leading seller of a related product in the same geographic market may also raise serious risks of tying. This problem arises because banks tend to enjoy a significant degree of monopoly power in local markets. Such power rests on a number of considerations. Banks are protected by regulation from both free entry and full competition. Also, banks enjoy power *vis-a-vis* borrowers because the borrowing relationship tends to be a continuing one, with little effective opportunity for "shopping around" for most borrowers, and such a relationship requires considerable disclosure of confidential internal business information necessary for credit evaluation. In times of tight money, banks "ration" a scarce and essential commodity to "good" customers. When

banks "ration" credit (or are forced to ration it) rather than raising the price to the point that supply and demand meet, *the process produces unexercised monopoly power—power which must be abandoned or used through transfer into some other market. It can be done by tying. No coercion is necessarily involved.* Rather, the borrower, recognizing the bank's discretion and economic power, is tempted *voluntarily* to patronize bank-affiliated enterprises in the hope of improving his chances to obtain it on favorable terms. *In antitrust parlance, this is now called "tying effect"—to emphasize that the economic effect is just as serious as with overt tying.* [Emphasis added]

But the law is slow and the computer fast. It has taken since the Sherman Act was passed in 1890 to develop current antitrust law. Some of our antitrust precedents date back to earlier common-law days, but we don't yet have clarity in many areas—including those with respect to reciprocity and cross-marketing. The computer industry even has its own much more current and immediate example of judicial delay.

The computer-services industry's trade association, Association of Data Processing Service Organizations (ADAPSO), has been concerned with the incremental-marketing issue for a decade, but its labors are only just now beginning to bear fruit by recognition of the problem of "tying effect" and by favorable administrative decisions, such as the FCC's limitations on computer-services incremental marketing by regulated communications carriers.

In 1967, following futile efforts to obtain relief from (or even a meeting to discuss the subject with) the Comptroller of the Currency, who had jurisdiction with regard to the incremental marketing of computer services by national banks, ADAPSO filed suit in federal court in Minneapolis. It sought a declaratory judgment that the public marketing of computer services by national banks was unlawful and requested injunctive relief prohibiting such activity in the future.

This was and still is a new area of the law, and in filing suit it was known that there were all sorts of dilatory legal defenses which might be asserted. But ADAPSO was also convinced that the issues were of sufficient importance to justify the struggle, and hoped that ultimately the U.S. Supreme Court would apply precedent with a view to the great changes the computer age was introducing, rather than in the sometimes stricter fashion of lower courts. In this latter hope ADAPSO was proved right, although it turned out to be a Pyrrhic victory.

The first defense motion was to dismiss the case without considering the legal or economic merits at all, on the grounds that ADAPSO technically had no "standing to sue." The motion was extensively briefed and argued. The Minneapolis District granted the motion and dismissed the case on January 9, 1968. The Federal Court of Appeals for the Eighth Circuit in St. Louis affirmed the dismissal on February 6, 1969. A writ of *certiorari,* the applicable form of U.S. Supreme Court review in this kind of case, was applied for and granted. Finally on March 3, 1970 the Supreme Court reversed the two lower courts and sent the case back to the Minneapolis court to begin again the lengthy process of pretrial proceedings and trial on the true merits—at the end of which undoubtedly the losing party would begin the review process all over again. Three years had already elapsed.

ADAPSO v. *Comptroller* in the Supreme Court is frequently cited by lawyers because of its landmark ruling enlarging the concept of who has "standing to sue." It has come to have especially great significance in environmental law controversies, where the complaining parties are frequently individuals and organizations only very remotely connected with the particular matter at issue. The case, in which I had been one of ADAPSO's counsel, was even cited *against* me in a litigation now pending before the Environmental Protection Agency which seeks to balance the risk/benefit equation with respect to use of the herbicide 2,4,5-T.

But *ADAPSO* v. *Comptroller* was not a victory for the computer-services industry. During the four-year procedural fight the banks had mounted a massive legislative effort and persuaded Congress to enact the Bank Holding Company Act of 1970. That Act in substance shifted the bank incremental-marketing battleground to the Federal Reserve Board and for practical purposes mooted the pending case against the Comptroller. ADAPSO accordingly dropped its Minneapolis litigation, although it continued the struggle against bank incremental marketing before the FRB and elsewhere. At the end of 1973 it even began another lawsuit, this time charging the Federal Home Loan Bank Board (FHLBB) with unlawful computer-marketing activities.

In contrast to the delays which are so characteristic of the progress of the law, the computer age is moving at computer speed. It was only the second Friday in April, 1946, when Dr. J. Presper Eckert, Jr. and Dr. John W. Mauchly at the University of Pennsylvania's Moore School of Engineering are said to have completed the soldering of some 500,000 connections linking over 18,000 vacuum tubes together to complete the first electronic digital computer in history, ENIAC (*e*lectronic *n*umerical *i*ntegrator *a*nd *c*alculator). Today's computers have no tubes (we're three computer generations later, at large-scale integration, having passed through the vacuum-tube, solid-state and integrated-circuit generations), are a small fraction of its size, and produce "throughput" (results) at costs which are several orders of magnitude lower. Although the computer is already so commercially embedded in our way of life that we couldn't live as we do without it, its significant commercial use is barely 15 years old. The chances are good that if we wait for computer-created economic issues to ripen into justiciable legal controversies and then be decided in leisurely judicial fashion, the camel carrying computer related problems will be within the tent before the law has decided to keep it out, at which point we will have to live with whatever has happened.

Solving the data-bank problem involves the judgmental balancing of conflicting interests of such massive importance to mankind as to call for a major public effort over a long period. Solving the economic problems presented by the computer seems by comparison a far simpler task, requiring primarily that the issues be recognized and appreciated for what they are. Once understood, the solutions are *comparatively* (although not absolutely) easy.

For example, an effort to obtain the benefits of POS marketing for all without introducing rigidity into the structure—in fact by creating flexibility and greater choice between alternative suppliers and therefore more rigorous competition—might at least begin by requiring that the interface between independent supplier and customer be such that one could switch from supplier to supplier exactly as one switches from one electric wall outlet to another. Once standards permitting such interconnection existed, one might be able to select alternative suppliers perhaps even more easily than one does today.

The American Iron and Steel Institute announced its Steel Customer Communication System (SCCS) designed to standardize nomenclature for all steel products so that customers will use a uniform product description when ordering. In early March, 1974, it developed a new computer-language code, called COMPORD, for computer ordering, which translates SCCS into computer language.

The grocery industry has already adopted a Universal Products Code—small vertical bars—which identifies shelf items and their manufacturers. The objective is that by the end of 1975 substantially all prepackaged foods will carry Code symbols preprinted by the manufacturers.

The National Retail Merchants Association (NRMA) is also proposing a set of standards which it hopes will be adopted for merchandise coding and credit-card processing. Some 40 billion retail merchandise tickets are now being used each year, with 60 billion anticipated before 1979. NRMA's merchandise-identification task force carefully studied three approaches to automatic scanning: optical font, black and white linear bar code, and magnetic coding. In December, 1973 its Systems Specifications Working Committee endorsed *o*ptical *c*haracter *r*ecognition (OCR) as the most desirable technology for use in merchandise and customer identification. The Committee's report stated that the Committee recognized the need for further evaluation and that efforts to develop a satisfactory functional optical-font technology and an acceptable OCR font would require industry cooperation. One of the officers of NRMA's Information Systems Division stated:

> NRMA has coordinated the project and enlisted the assistance of manufacturers and suppliers, as well as the Depart-

ment of Commerce, National Bureau of Standards, in order to obtain a compatible technology for use in the future.

If these kinds of goals can be achieved, it would see that more generalized POS standards are equally possible; MICR is certainly proof that a trade association can enforce standards.

The development of standards for POS or any other computer application is certainly not a simple matter. Computer technology continues to develop rapidly. Computer hardware (the physical equipment) and especially the main frame (or central computing part) is sufficiently advanced in speed of operation and reduction of size to be able to handle our commerical needs for some long time to come. Although the sophistication of a few software systems (the instructions to the computer as to how it should function) may have kept pace with that of hardware, many computer scientists believe that software is still relatively rudimentary. Advocates of such computer languages as BASIC and APL claim to be able to impart a working knowledge of their languages to an intelligent prospective user in one hour, but useful software still requires major breakthrough before the ordinary individual will be able to communicate easily with the computer. Freezing computer developments at too early a time may thus inhibit progress and restrict competitive innovation. Freezing the width between railroad tracks in the mid-nineteenth century may have prevented the development of better balanced or more economical railroad cars. And freezing the male electric plug at two prongs has certainly prevented the general introduction of the safer and preferable three-pronged plug with ground wire included. Moreover, standards necessarily limit to some extent the distinctions that form the basis for competition between computer systems. If all computer designs (the most significant of which are now mainly software) are identical, industry enthusiasm may be dampened so as to limit the market. In retailing, as in banking, users want to have an "edge" on their competitors; they don't want to be undistinguishable except where cost benefits are paramount.

But it is possible in developing standards to differentiate in some degree between computer architecture and implementation;

one could freeze the *specifications* of a particular computer, but not its semiconductor or other *technology*. And standards can be upgraded to include compatibility with previous standards.

Difficult as the choice may be, the development and adoption of POS standards is a feasible one and is in fact already upon us as a matter of commercial necessity, although now being decided by private rather than public interests. Thus, in contrast to the individual association efforts referred to above, in late 1972 after long study and research, two independent groups, the American Bankers Association (ABA) and the International Air Transport Association (IATA), decided upon a standard configuration for the magnetic stripe on the back of credit cards. Although their decision doesn't have the force of governmental law and anyone is free to develop an alternative, the chances are excellent that in due course their proposal (and NRMA's and others when widely used) will become the *de facto* standard which all must follow. Can anyone imagine offering a magnetic encoding device to banks which will print magnetic numbers on checks in a different place or of a different size than MICR checks now contain?

Unfortunately the government has left most of the development of standards to private industry and organizations such as ABA, ANSI (American National Standards Institute), IATA, NRMA, and the Steel Institute. Indeed in many respects IBM's size and position of leadership in the computer industry is such that its developments of both hardware and software constitute standards which others follow. (This tendency for IBM products to become *de facto* standards should not be confused with the above deliberate process of standardization, because of the legal implications involved, but the economic consequences in many respects are much the same.) The National Bureau of Standards does set standards for certain government transactions, although not all. Otherwise its authority is only advisory. And—perhaps partly due to lack of funding—even where it does function, its activities are unfortunately most limited.

Standards are a commercial essential. If they are to exist, they should be determined by democratically selected public judgments and not because of the special leadership position of a single private company or by groups of such private companies agreeing with each other. The latter in fact smacks of conspiracy in restraint of trade and may well be unlawful. Recognition of a further computer principle is required:

> **Standards:** *Computer standards should be fixed by fairly selected and representative public organizations, so as to encourage maximum reasonable interchange among computer systems and between economic units, without unreasonably impeding technological development.*

Just as properly promulgated and regulated standards might solve the POS problem, early recognition of the incremental-marketing issue and aggressive antitrust enforcement by the FTC and Antitrust Division and by other agencies with similar kinds of jurisdiction, such as the FRB, the FHLBB, and the Comptroller's office, would control and perhaps eliminate that problem as well. But unfortunately up to now the FCC is the only agency which has seen the light—although a late 1973 Federal Register notice by the FRB soliciting comments regarding the checkless/cashless society is a hopeful sign that it may be studying the matter. Few antitrust experts seem even to recognize the existence of the issue.

Here, as elsewhere, the evidence continues to mount that what is really needed is public understanding of what the computer is doing and of its consequences. We are lulled into a kind of drugged sleep by the tremendous promise the computer offers us of traveling to the moon and planets and curing disease and poverty, unaware that the price we may pay could be our heritage of economic freedom.

6 SECURING THE ESTABLISHMENT

LEXIS is a computerized legal research service. It is now operational and being offered in New York and is planned for gradual introduction nationwide within the near future. A predecessor version has been available for some time in Ohio. To appreciate the tremendous benefits this computer utility offers requires some understanding of American common law and traditional legal research methods.

The federal and state constitutions and statutes rarely cover a precise point at issue in sufficient detail to remove all doubt. Many of these documents were carefully written in general terms, expressing broad principles to be applied on a case-by-case basis over the years as experience developed, and leaving as much flexibility as possible to permit future generations to deal with future problems.

Where constitutions and statutes are not directly applicable, precedent is found in the "common law," which is the body of case experience going back to our English heritage. The great bulk of legal authority is therefore found in cases decided by the courts, interpreting and construing constitutions and statutes in specific situations, or expressing the common law.

Whenever a lawyer is presented with a problem to which he does not know the answer, he must find whether there is a constitutional or statutory provision in point. He must also usually determine whether there are cases construing the applicable provision which will be helpful in solving the problem by analogy, in his own state or elsewhere. Rarely is there even a controlling constitutional or statutory provision that doesn't require some kind of interpretation or construction. As a result, many lawyers often turn first to the case law, and work their way backward from cases to statutes and constitutions if necessary.

Although researching the cases is a task often awarded to the youngest lawyer, proper research can be a time-consuming, backbreaking job entailing a great amount of time and therefore tremendous expense to the client, for lawyers usually charge, directly or indirectly, on the basis of the hours they work on a matter. Good legal research requires finding useful precedents among millions of federal, state, and local authorities over the years, as well as reviewing the efforts of scholars who have written on the subject here and in other common-law jurisdictions such as England or, on occasion, even in civil-law countries such as France.

There are a number of different tools available to the lawyer to help him find the statutes and cases in point, and any good research job in a difficult area usually requires use of several of them. They include encyclopedias, digests, annotated statutory provisions, citators, and a great variety of law services. But sooner or later in one way or another they all come down to a "subject matter" approach. This means that someone classifies the statute or case in one or several categories, and the lawyer finds it by looking in that category. Obviously judgments on the parts of both classifier and researcher are involved, and sometimes their decisions as to the applicable category will differ so that the case is not located. For this reason, complete research usually requires using several of these tools, on the theory that at least one of the classifiers will apply the same judgment as the researcher, and the case will turn up. Even here, however, a case is occasionally missed, and most lawyers heave a sigh of relief when they read their opponent's brief in an important matter and find they have not missed a significant decision.

Securing the Establishment

The expense and difficulty of performing legal research is one important reason why the impoverished or otherwise deprived so often have something less than first-class legal advice and counsel. Many clients just can't afford to pay for the time required, and their lawyers as a result have gotten into the habit of doing little or no research and playing even criminal cases "by ear" or "out of their hats."

LEXIS is a completely new approach made possible by the computer alone. It is a "full text" research system, by which is meant that every word, every comma, every paragraph, and every other part of the constitution, statute, regulation, opinion, or case is reproduced exactly in the computer's data bank, without editing, digesting, or modification in any form. The data bank is queried by a typewriter-like keyboard through which the lawyer interacts with the computer until he gets his answer; the answers appear on a television-type screen. For example, if he represents a client whose infant child has been killed as the result of a fall down an elevator shaft because the door was left open, he doesn't have to apply such legal terms as "attractive nuisance," "negligence," or "tort." He simply asks the computer to display the citations for all cases in which the phrases "elevator shaft" and "fall" appear. If there are too many, he can add the words "death" and "child" or "infant." If there are still too many, he can limit the search to cases in the Supreme Court of his state, or where damages of over $10,000 were awarded, or which were written by the particular judge before whom the case will be tried. If the lawyer wants a printed copy of the results of his search, he presses a button and the writing on the screen is printed automatically.

LEXIS makes possible some kinds of research which are ordinarily impractical with present methods. All cases involving a particular disease or injury can be collected almost instantaneously, where in the past no classification of this kind existed. The decisions of a single judge in a certain type of case can be collected to show the judge's inclination; a judge's attitude can sometimes be the key to success in a case. Similarly, all of the cases in which an adversary attorney appeared as counsel can be obtained by simply entering that lawyer's name into the terminal. (The "terminal" is the keyboard, television-type screen, printer, and associated equipment which is remote from the computer itself.)

At least for the present, LEXIS is only an addition to the available research tools, not a replacement. But it is an enormously valuable addition which should both improve the quality of legal services and drastically reduce the time required for research, making possible lower legal charges. Because LEXIS requires a minimum of legal background and training, some lawyers already use paraprofessionals (specially trained nonlawyers) to operate their computer terminals, thus reducing legal-research costs even further. More will undoubtedly follow.

Many of us consider that the constitutional right to counsel is infringed every day in our criminal courts, where defendants are assigned (or sometimes unwisely choose) attorneys who are unqualified except in title and license to practice. LEXIS may not be the complete answer, but it certainly has the potential to reduce the inequality between research tools available to the rich and poor lawyer, and between legal services available to the rich and poor client. The outcome of controversies and prosecutions can be made to turn more upon the merits of the case itself and less upon legal representation.

But—LEXIS is presently being offered on a "charter" basis to a limited number of users who sign long-term contracts committing themselves to a minimum direct out-of-pocket payment to LEXIS of 36 thousand dollars a year. At least an additional equal amount of indirect overhead costs must be contemplated for such items as attorney and staff training and the like. Only the wealthiest law firms, ones which ordinarily represent large and powerful corporate clients, can afford this kind of commitment. It is a safe bet that few law firms or attorneys in Harlem and Watts will sign up. Furthermore, the manner in which LEXIS is being developed compounds its exclusivity and guarantees that few such firms would sign up even if they could somehow raise the money. This is because its initial legal data base is actually being designed to serve the needs of the wealthy corporate practitioner rather than of the criminal, civil-rights, landlord-tenant, negligence, workmen's-compensation or poverty attorney.

The total volume of federal, state and local constitution, statute, case, and regulatory law and interpretation is massive and will take years to place in computer memory in full text, even going full blast. A start must be made somewhere, which of course means a selection. LEXIS and its sponsoring bar associations are going about this by the *economically* logical method of selecting materials geared to the needs of the anticipated initial large users, using a committee advisory structure controlled by such users. Little wonder that the first focus of interest is on tax, SEC, trade regulation, and corporate materials, rather than on such individually smaller but collectively far more important matters as automobile negligence or workmen's compensation cases. Yet it is in the latter areas in which the system could find its greatest utility. No other existing legal-research system, for example, would permit an attorney to uncover quickly all workmen's compensation cases involving a specific type of physical injury ("herniated, intervertebral, nucleosis pulposis"), which would be of major precedential value in setting damages. Some of these kinds of cases do get into the system, of course, but only if they happen to be included within categories selected for other reasons. One cannot rely on their completeness.

LEXIS is a private commercial enterprise. It is understandable that an entrepreneur which will have invested perhaps twenty million dollars before it reaches break-even, will choose a development and promotional plan likely to ensure financial success to itself and result in a return to its stockholders and investors. Social considerations necessarily take a back seat. LEXIS' success, however, if it comes, will be the result of a cooperative effort between it and the organized bar associations of the states in which it operates. Bar associations are public or quasipublic organizations. Each will for practical purposes have granted LEXIS the substantial equivalent of an exclusive franchise to work with attorneys. Even judges may be persuaded to turn out their opinions in computer-readable form as a part of this cooperative program—with a special type of ball on a Selectric typewriter they would already have machine-readable text.

Under present plans and despite some publicly available

terminals, it therefore seems probable that for the first few years of its operation LEXIS (as a service for lawyers) will be available primarily only to the large law firms which represent wealthy corporate clients and individuals—the "Establishment." Whatever the ultimate consequences of this development—and one hopes they will be to reduce costs and make better legal services available to all—at least the short-term effect of the introduction of LEXIS (assuming its success) may be to *widen* the disparity between legal services available to rich and poor, Establishment and deprived, not to narrow it.

This is one of the great problems presented by the computer. It is extraordinarily effective and efficient, and can dramatically reduce costs and increase quality. But, as we earlier pointed out and is worth repetition, it is tremendously expensive to install and make operational. Consider who has the giant computer and the advanced applications today. It is the huge utility, the automobile manufacturer, the large insurance company, and the organizations represented by the law firms with LEXIS itself, not the corner drug store or the small clothing manufacturer. True, these smaller companies may have minicomputers or access to independent data-processing centers to produce routine accounting services such as payrolls or accounts-receivable processing. In comparison with what is available to those who can afford it, however, these and the few LEXIS terminals which may be available to the bar at large represent little more than tokenism.

The computer's reinforcement of the Establishment represents one of its serious problems to society. There are no easy solutions. But one thing is certain—there will be *no* solutions if we don't think about the problem and what is happening before our time runs out.

LEXIS is just one example of the tendency of the computer to reinforce the Establishment, giving it greater advantage than ever. POS marketing is another. Despite its potential to reduce economic entry barriers to new vendors, POS will almost certainly be available initially only to the great chains. Almost any other advanced computer system is likely to function in exactly the same way. These latter instances, however, are characteristic of our modern industrial society and symptomatic of the economies

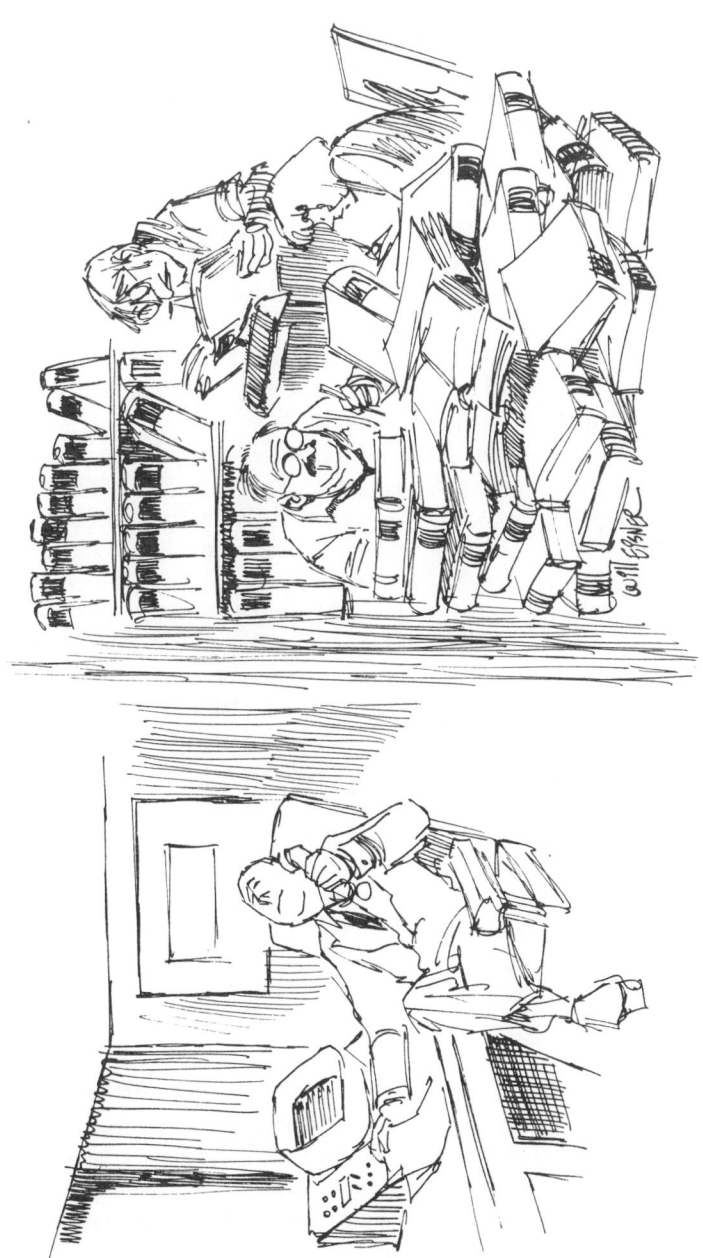

of large-scale operations—General Motors will probably always be first to have the new giant metal press and General Foods the new multimillion-dollar packaging machine. But LEXIS is different, because its reinforcement of the Establishment is largely unnecessary, and might well be avoidable were there understanding of its effect, and public pressure for more democratic distribution of its benefits. LEXIS is in fact a particularly persuasive example, because the edge it gives the favored few is in an area subject to considerable direction and control by the organized bar itself— the very segment of our society which presumptively should be most alert and sensitive to all this. In this respect it demonstrates once again the inadequacies of society's understanding of the computer and its impact.

LEXIS is being marketed as a money-maker to lawyers, and it appears that law firms using its predecessor service in Ohio have earned profits with it. Ingenious financing, including bar association self-assessment and legislative fundings, accordingly come to mind as possible alternative programs by which its service might be more equitably distributed among lawyers and others; the bar has been most successful in funding and constructing huge bar centers and other buildings and institutions. Before this can be expected to happen to LEXIS, however, there must be recognition by its sponsoring governmental and quasi-governmental bodies, of their obligations to supply computer services in a manner which will serve the public generally, and not unfairly benefit any special segment—the further computer principle:

Public Computer Service: Public and quasipublic-sponsored computer services must be supplied on terms and conditions which result in their fair and equitable distribution to the public.

The design of a new and equitably distributed program for LEXIS from its inception in any state would undoubtedly require a tremendous effort, as well as considerable creativity and ingenuity. That has not stopped lawyers in the past, from Magna Carta to our federal Constitution. It should not stop them now.

7 UNTYING THE COMPUTER GRID

At exactly 5:16:11 P.M. EST. on November 9, 1965, an electromagnetic relay which had been set too low at the Ontario Hydroelectric Commission's Sir Adam Beck Plant No. 2 near Niagara Falls, New York, broke its circuit even when the power fluctuated within normal limits. For some yet-to-be-unraveled reason, when the relay opened the circuit breaker on its own cable, that cable's current jumped to the other five cables of the six-cable power line, knocking them out as well.

The Beck plant was part of the northeast power grid, called the Canada-U.S. Eastern Interconnection (CANUSE). CANUSE interconnected electric utilities all over the east, permitting them to shift power back and forth as required. Immediately the surges of excess power moved through the whole grid, and within 16 minutes 30 million people in an 80,000 square-mile section of the northeastern U.S. and southern Canada were shrouded in darkness. People were trapped in elevators and subways, commuters were unable to reach their homes, and airports were closed.

Fortunately the weather was clear and not too cold, and the damage, though great, was not catastrophic. People were thankful that what might have been did not happen, a lesson was learned that the authorities and utilities seem to have corrected, and life has gone on.

91

But the lesson was *not* learned in its application to at least one other area. The automated payments mechanism we are creating to link banks and other establishments throughout the nation will tie these economic units to each other in as intimate an electronic fashion as any public utilities in a grid were ever connected. Point-of-sale marketing will link units of the distribution chain in an equally tight manner. The already operational computer utilities tie their economic segments together so that the individual parts are usually unable to operate independently. The electronically interconnected activities of all these computer applications are efficient *because* of their economic interrelationship with each other and with the computer, and simply cannot function alone; mechanical or human substitutes are not possible.

It may well be that generalized *power* blackout problems have been solved, and that the nationwide bank interconnections cannot be turned off in that fashion. But once we have tied together the various components of our economic system into a single interrelated whole, ignoring the possibility of *economic* blackout as the result of negligence, error, strike, sabotage, natural calamity, war, or otherwise, we invite a repetition of the 1965 disaster, this time on a far more serious level.

The independent components of any nationwide economic computer grid may each be private companies when separate, entitled to their own successes and failures and to make their own mistakes. But once they join together, society must be given the right to demand freedom from the adverse consequences of their joinder. If an electronic system tying together retailers, clearing houses, banks, credit companies, and others all over the country goes down, how does the Chicago retailer collect his bills—or even find out the status of his accounts? Nor does it much help the electronically interconnected small businessman who cannot get supplies or transact his affairs, to be able to assert a claim against his vendor if the latter is bankrupt or has the excuse of national disaster or sabotage.

In the well-designed computer installation, an effort is made to bypass problems by providing redundant resources or alternative paths among resources. The related problem of protecting a particular computer installation from its users is also well-

known, at least at university computer centers. Means of coping with the situation are quite effective when carefully established. But these security measures are not designed to deal with the failure of an economic computer grid, and the same kind of attention has not yet been paid to that larger issue. Uncontrolled and unregulated, damage to any part of such a grid can have a domino effect which must be considered socially unacceptable. Avoiding that effect should not be a major problem and need not involve any new or super-governmental regulatory agency such as the Federal Power Commission, for the computer grid is not the same as the electric power grid; it will develop as an ultimately integrated economic whole, if at all, only because its existence is not questioned and limited in the first place.

Here, as in so many other computer applications, what is needed is primarily an appreciation of the problem. Once recognized, the solution to catastrophe almost jumps forth:

> ***The Computer Economic Grid:*** *The failure of a discrete unit of a computer economic grid must result in immediate disconnect from the grid without unreasonable harm to or interference with the rest of the system.*

Yet obvious as this may be, it is not yet even a recommended official standard (because there are none!), much less any requirement of law or regulation.

COMPUTER-RELATED CRIME

Control of the computer grid's domino effect must contemplate avoiding deliberately induced failure as well as negligence or unavoidable error, for computer-related crime will soon be here. Indeed, most future major commercial crimes will undoubtedly require some degree of computer involvement. Although apparently not a computer-related crime as such, for example, it is clear that the recent massive Equity Funding pyramiding fraud would not have taken place without computer involvement.

Prosecutions have already resulted from the breaking of secret codes or passwords and the ensuing tapping into remote time-sharing systems to acquire valuable confidential information and services. The first federal criminal case involving the use of a computer occurred in 1966. A young programmer working for a software firm was responsible for the programming and occasional operation of an IBM 1401 computer at a bank where he also had his checking account. He placed a change in the program to ignore his account when checking for overdrawn accounts in exception reporting. His plan was to leave the change in the program for only three days during which he knew his account was going to be overdrawn by $300. Four months later his account was still overdrawn by $1352. The change was still in the program, until fortuitously and fortunately the computer failed one day, resulting in manual processing which uncovered the overdraft.

One story, yet to be traced to its source and possibly apocryphal, tells of the bank computer programmer who modified all the bank's programs to cause the odd penny "breakages" in computations of interest and other bank charges to customers all to be placed in a single special account to his credit. No single customer would recognize or care about the loss of less than a penny in any transaction, but the total amounted to many thousands of dollars. Another story tells of the man who printed up a vast number of blank bank-deposit slips identical in printed appearance to the bank forms, but with his own account number MICR-encoded invisibly in place of the special bank identification for such slips. He placed them in all the bank's branches one day, let the deposits accumulate to his own account, and then withdrew the total and fled to points unknown with the cash before any depositor learned of the switch. Crimes of this kind in a computer grid environment could have massive economic consequences.

Donn B. Parker, a computer scientist at Stanford Research Institute, has done a fine job organizing and assembling the evidence of computer-related crime, and has worked long and hard to try to alert society to the problem. His work includes both crime in which criminals attack honest computer systems, as in the case of the cheating programmer, and that in which criminals establish dishonest systems, as in Equity Funding. Incomprehensible and esoteric as these types of conduct may seem to the layman, Parker's work suggests that they should not be as difficult to avoid as the typical white-collar crime, or crime of avarice, passion, or prejudice. Once the possibilities are recognized, it is within present technical know-how and not necessarily very difficult to build the safeguards right into the computer system itself. For example, any unusual level of payments or deposits in an account can be flagged and stopped until confirmed, and transactions of any other special kind can be held up pending approval. I recall outlining the crime potential in a talk in late 1972 to the Japanese Institute of International Business Law in Tokyo. Those present hadn't experienced the problem yet or thought about it very much. But the moment they appreciated the possibilities, they promptly indicated they would take steps to introduce the obvious controls. IBM has funded a 40-million-dollar long-term study of both physical and logical security which will attempt to develop major new recommendations in this area and render it more controllable. The long-range problems seem clearly to be social ones, not technical ones.

As in so many other respects, the initial key to computer grid control is recognition of the serious consequences of grid failure and appreciation that it may result from willful as well as negligent or unavoidable error. Once that is accomplished, other far more sophisticated and esoteric measures will undoubtedly be suggested by those computer professionals assigned the task of providing for security against grid breakdown. Surely one rule would be that only the minimum number of units required to accomplish the desired result should be linked together, with the guiding overall principle being that the consequences of any potential problem must be held well below the benefits clearly provided by the joinder.

8 THE VITAL HUMAN FACTOR

The human factor is the final unknown in the computer risk/benefit equation. It is made up of such tenuous intangibles as emotional and intellectual impact. Its analysis and evaluation may be the most difficult assignment of all, and in some respects it may turn out to be the most urgent.

HUMAN RESPONSE

Our home is heated by gas. Two years or so ago our gas bills began to rise dramatically, and I was convinced that something was wrong. Either the gas meter was recording inaccurately, or there was a leak someplace, or Con Ed's records were somehow confused.

I called the sales representative who had handled our changeover from oil to gas some time before, but he was sick and no one else seemed to know much about it. The representative who finally came to our home reported that all was well. Still, the consumption continued at the increased level. I wrote several letters to the company in an effort to get some satisfaction, but they went unanswered.

Finally I complained to New York State's Public Service Commission, which regulates gas companies and which has an efficient Testing and Consumer Section to handle customer complaints of this kind. They quite promptly sent an inspector, who couldn't find any trouble either, but did discover that Con Ed hadn't been around to read our meter for a long time and had been estimating consumption. He read the meter, found a substantial difference from the estimated figure and put through a revision to Con Ed. I received a second bill during the middle of our regular monthly computer-billing period, based on a new and now current reading but issuing no credit even though I had already received and paid the regular monthly bill. Apparently there was no easy way to handle this kind of interstitial billing. As a result, days after I paid the amount I actually owed rather than the amount shown on the new bill, I received a special computerized form in red, entitled "TURN OFF NOTICE," announcing that if I didn't pay within ten days, my gas would be turned off. I paid. The next computerized bill came soon after, this time duplicating a part of the earlier readings. I tried to reach someone with whom to speak, still to no avail, and finally paid again.

I am a meticulous record-keeper and attender to financial details, perhaps excessively so. All ultimately worked out properly, but my irritation at being required to pay what I knew to be duplicate and incorrect invoices or risk automatic penalty of no heat, hot water, or cooking facilities for my wife and children was very real. What was especially frustrating was having absolutely no one to speak to or complain to except a machine, which would only belch back to me, "pay up or no gas." Had I not been a trial lawyer with knowledge of the burden and difficulty of going to court, I might well have sued Con Ed.

My experience is certainly not unique. In a recent Ohio case, a number of residential customers of natural gas supplied by Columbia Gas Company brought a class action for injunctive and declaratory relief and for damages, alleging that their gas service had been terminated under color of state law in violation of their constitutional right to due process.

The Columbia Gas Company is a large, privately owned, pervasively regulated public utility company. It serves over 140,000 customers in the Toledo, Ohio area, and all of its billing is handled by computer in Columbus, Ohio. A reading is normally

taken from each customer's meter every second month, although (in a fashion identical to my case with Con Edison) on occasion no reading may be taken for a period of many months. When no reading is taken to reflect actual usage, the company's computer estimates usage and calculates an amount which is then billed to the customer. For some reason which is not made clear in the court record, Columbia's computer usually underestimates in these situations; consequently when an actual reading is eventually made after a series of several computer estimates, the resulting bill for actual gas consumed can be surprisingly high. In those cases the customer, especially if poor, often is unable to pay a bill several times larger than normal.

Whenever a monthly bill is not paid by the customer, the amount is carried forward and added to the customer's next bill. If the second bill is not paid by five days after the due date and the amount in arrears is $20 or more, a notice of termination (a "shut-off notice," just like my TURN OFF NOTICE) is sent to the customer. The practice complained of was at the next stage, which fortunately I never reached—if payment was not made within five days of the issuance of the shut-off notice, an employee of the company went to the residence and terminated the service. Although this employee was authorized to grant temporary extensions of time in which payment could be made, he was under no obligation to inform the occupants of the premises that he was about to terminate gas service. He usually made no contact at all with the occupants, even to verify the correctness of the address.

The evidence established that however imperfect the company's procedure was in theory, in practice it was more so. "Significant and tragic mistakes" were often made; for example, one witness testified that his gas service was terminated even though he had paid his bill in full upon receipt of a final notice. One of the parties testified that his gas service was unexpectedly terminated on January 4, even though he had paid his bill by mail on December 30. When he contacted the company by telephone and informed them that he had paid the bill, an employee of the company replied, "Tough. Pay the bill again." This customer had seven children, and the temperature in his house dropped to 45 degrees before service was eventually restored through the intervention of the Board of Community Relations.

When a customer made special arrangements for deferred payments of a larger-than-usual bill, the monthly statement would be accompanied by the form shut-off notice, which the customer would be instructed to disregard. One party testified that after having been billed about $12 a month for a series of estimated bills, she received a bill for actual usage for over $197. She made special arrangements to pay this large amount over a period of months, during which time she received a shut-off notice each month and an additional notice requesting that she disregard the shut-off notice. Despite this, and although she paid the stipulated amount monthly, her service was terminated in mid-December; it was not until the eventual intervention of her church pastor that the company finally acknowledged its mistake, apologized, and restored service.

Administrative and clerical errors also resulted in unpleasant surprises for the company's customers. One witness testified that he received on December 30 a notice that his service would be terminated if his bill were not paid by January 4. He too mailed his personal check to the company on December 30, which was endorsed by the company and cashed on January 3. Nevertheless, on January 4 his service was terminated. The first that he, his three children, and his pregnant wife learned of the termination was when it started to get cold in the house. The company showed that a clerical employee had misplaced the record of the customer's payment, thereby causing this unwarranted termination.

After two days of hearing what another court aptly described as a bizarre "Orwellian nightmare," the lower federal court concluded:

> The evidence as a whole revealed a rather shockingly callous and impersonal attitude upon the part of the defendant, which relied uncritically upon its computer, located in a distant city, and the far-from-infallible clerks who served it, and paid no attention to the notorious uncertainties of the postal service.

This case was clearly an extreme one, and the court granted relief requiring personal notification of termination, opportunity for protest, and a sensible method of resolving disputes. Most cases, however, do not reach this level and the frustration continues. Indeed, such human frustration in dealing with a computer system is already so sufficiently generalized that a number of credit-card companies now identify on their bills a person with whom the customer can speak if he has a problem. A great idea, perhaps, but the times I've tried to contact that person I've found he doesn't exist but is just a slot on a PBX. He's a "nonperson"—an "unperson." Instead of being eliminated, the frustration then becomes compounded. Undoubtedly it was in part for this reason that New York recently enacted legislation requiring creditors to furnish prompt and specific response to complaints, at least before supplying unfavorable credit information to credit agencies. As a result, New York customers in late 1973 were deluged with form notices, such as the following from Avis Rent-A-Car:

NOTICE TO OUR CUSTOMERS

If you wish to allege that a billing error has occurred you are required to do the following:

A. You must send your objection in writing; and

B. The objection must be set forth in such a manner as to enable Avis to identify you and your account, the amount in dispute, the transaction shown on the statement or invoice which you believe to be in error, and the facts providing the basis for your belief that an error has been made; and

C. The writing must be sent by registered or certified mail, return receipt requested, on a paper other than the Avis statement or invoice and,

D. You must act within 30 days of receipt of the Avis invoice or statement.

Once you have complied with all the above steps (A thru D inclusive), Avis is required to do the following:

    I. Mail a written acknowledgement to you within 30 days of receipt of your inquiry.

> II. Within 90 days after receipt of your inquiry, and prior to taking collection action, either:
>
> a) Make appropriate corrections and mail a written notice informing you of the fact that the error has been corrected and will be shown on the next regular statement, or,
>
> b) Send a written notice to you setting forth the reasons why Avis believes your account was correctly shown on the statement; and
>
> III. Not communicate any unfavorable credit information based upon the disputed billing error to any person, including credit agencies, until Avis has complied with I and II above.

My secretary has had her own separate and lengthy battle with a Con Edison computer. She too received one of these forms. Her delight in thinking she might finally get someone to respond to her continuing pleas for correction of her bill was such a pleasure to behold that I didn't have the heart to tell her that the statute didn't apply to her problem!

Compared with some of the other problems the computer presents, the frustration people seem inevitably to experience in dealing with the computer which handles their bank statement, their credit card, telephone, other utility, and store bills may be a minor inconvenience reflecting just another one of those things which modern life requires. But as the computer becomes omnipresent, as surely it will, this area of concern will grow correspondingly. Here at least we should be permitted to enjoy the satisfaction of being the computer's master rather than its servant.

John T. Swarens was a freedom-loving Kentucky citizen who purchased a Ford automobile on credit from his local Ford dealer. His security agreement promising specified monthly payments was assigned by the dealer to the Ford Motor Credit Company, which handles this aspect of Ford's business.

Swarens was a decent man who paid his debts, and each month he sent a check to Ford as required. Somehow, however, the inevitable occurred, and his account wasn't credited with payment. Ford sent collection agents to see Swarens, demanding payment. He showed them his canceled checks, which of course satisfied them and they went away. The error still wasn't corrected, and the agents came back two months later when payment wasn't reported. Again Swarens showed them his canceled checks, again they were satisfied and went away, and again the error wasn't corrected. When they came back a third time another two months later, he had had enough. Instead of exhibiting his canceled checks, he showed them his shotgun. Again they went away, but this time they *weren't* satisfied. Ford seized the car for nonpayment of the note and sold it.

Swarens brought suit, and proved his case to a court and jury. They awarded him not only the value of his lost car, but 250 percent in punitive damages, a kind of criminal penalty imposed on Ford for its misconduct. The highest court of the State of Kentucky affirmed the award in dramatic language, saying:

> Ford explains that this whole incident occurred because of a mistake by a computer. Men feed data to a computer and men interpret the answer the computer spews forth. *In this computerized age, the law must require that men in the use of computerized data regard those with whom they are dealing as more important than a perforation on a card.* Trust in the infallibility of a computer is hardly a defense, when the opportunity to avoid the error is as apparent and repeated as was here presented. [Emphasis added]

Clearly the Kentucky Supreme Court was telling us that man has a right to human interaction—it was enunciating what may become another fundamental computer principle:

> **Human Response:** *The supplier of computer services to the public must afford the ultimate consumer reasonable human response and interaction, or be liable absolutely for error and harm done.*

The Wisconsin Supreme Court has given implied support to such a principle by referring to "the assumption that there will be sufficient time after the mechanical processes are completed for the human factor of judgment to be exercised." And the federal Court of Appeals for the Sixth Circuit seemed to be saying the same thing when it affirmed the Ohio District Court's decision against the Columbia Gas Company. Its words in holding that the gas company could not rely exclusively on its computer in its billing practice are most compelling:

> The highly computerized collection and termination practices of the company are governed by a singular corporate concern for efficiency and protection of assets. However, the Due Process clause was designed to protect the rights of the citizens from procedures which often serve more to insulate the state from individuals than to serve their needs. The Constitution recognizes higher values than speed and efficiency.

Error is inevitable in both human and computer action, but if industry chooses to rely on a computer system and not interpose human judgment between the computer's decisions and the human victim (as with the gas company and me), then industry must be forced to change its ways either by direct regulatory action or by the imposition of a sufficiently large penalty to force it to mend its ways.

A requirement that there be human response in these cases would of course also minimize the effects of computer error, wholly apart from its consequences in relieving frustration. Computer systems are undoubtedly far less error-prone than man, and their errors are almost always the result of an operator or programmer mistake. (Perhaps 75 percent or more of computer error is the result of a flaw in the design or analysis of the system or a program, and 24 percent is keypunch- or computer-operator error. Less than a small fraction of one percent is electronic error; thus it seems incongruous to run a multimillion-dollar system with a low-paid and untrained clerk.) Despite what some laymen consider to be infallibility, however, computer systems *do* make mistakes, whether because of miskeying a keypunch input or a programming bug, or deliberate misconduct or crime, or—in very rare cases—even because a piece of dust prevents the sensors from feeling the hole in a punch card. Computer error rate may be a small fraction of human error rate, but the trillions and more of computations involved inevitably result in a sufficient number of mistakes to require their processing as a part of the regular course of doing business. That handling must be made to include human judgment as well as sympathy.

Educational aptitude and achievement scores have tremendous significance in today's competitive scholastic environment for young people. A few points one way or the other on an examination can spell the difference between admission to medical or law school and frustration for the budding doctor or lawyer. A computer-programming error might result in test scores being reported slightly higher or lower for a single category of students, or to one or a few selected institutions. It could be caused by simple failure to debug a program adequately. Or it might even be caused deliberately by an aggressive and ambitious father working for the testing agency, who inserts a program "patch" reporting moderately increased scores to selected institutions for all students whose last names begin with the same letter as his own. The change may be much too small to be apparent to the university admissions official comparing scores or even to an unsuspicious parent should he see it, yet be of major consequence to the young applicant and his future, and to others competing against him for the limited vacancies.

Errors, even of this kind, are relatively more extensive and more difficult to control in a manual system. But perhaps for those very reasons, most good manual systems have developed effective measures for control, and recognize the importance of responding to the questioning public in a manner which will satisfy the natural interest. To this point, that has not been true of the computerized system.

Important as is the language of the Kentucky Supreme Court in the *Swarens* case, it has not yet been cited or referred to by any other court in the United States or elsewhere—including even the Columbia Gas courts which dealt with the same kind of problem. The issue isn't adequately indexed in a classifica-

tion in which judges and attorneys with similar problems will search. Only a very few writers and observers of the computer scene have noted it. This is therefore still another reflection of the general lack of public information (and corresponding concern) about what the computer is doing—a lack which has thus far contributed so much to our failure to develop the necessary tools to manage the computer for society's total benefit.

HUMAN OBSOLESCENCE

Events move fast in the computer industry. As in other modern hard-science areas, the computer scientist who earns his doctorate and begins to work developing a laser or bubble memory or designing an advanced software system sometimes finds that he is far more skilled in the new technology than is his superior. If he is good, he moves up the ladder to administrative and executive responsibility. In the process he too may soon find himself out of touch with the rapidly changing technology. Because he cannot really understand and guide and control what his newly assigned subordinates are doing, he is now no more able to administer and execute than was his predecessor. (Computer people cite the aphorism, "Good programmers become bad managers.") And reeducation, although certainly possible, presents great problems. In the computer industry it often means "doing," and it is difficult for a senior employee to step down to junior level. As a result, in a very real sense the computer scientist may become "obsolete" and unable to perform his assignment.

Human technological obsolescence is not a new phenomenon, nor is it characteristic only of the computer industry. It was with us long before the commencement of the atomic and solid-state electronic ages. But data processing is already our eighth biggest industry, and (with its related communications) will surely be the largest by far within the next generation. For the foreseeable future, technological developments rather than stability and inactivity will continue to be the order of the day; thus the number of senior people obsoleted will become increasingly larger, and potentially important contributions to society by virtue of advanced education and abilities will be lost. Welfare programs, early retirement, and enforced leisure are not satisfactory solutions for the 35-year-old "over-the-hill" mathematician.

The technological obsolescence of at least some of these computer professionals can be turned to society's advantage and not lost, however, if we only will recognize the need for such trained leaders in dealing with computer/society issues. One could not find a better equipped group from which to select persons to educate the public in computer matters; to man the computer staffs and boards of the universities and federal, state, local, and civic agencies; and to conduct the broad interdisciplinary exchanges with the other professions and the public at large—functions which are essential to effective understanding and control of the computer and to the channeling of its efforts for mankind's welfare. Even though this is surely but a limited and partial solution to the problem of human obsolescence in the computer industry, and much more remains to be done, it represents an important and essential beginning.

COMPUTER INTELLECTUAL PROPERTY

The computer is quite capable of producing new, different, and very exciting forms of "art" and "music"; ultimately it will probably provide poetry and literature, as well. Hanging in my office are two award-winning color computer prints, "Hummingbird" and "Inspiration." Although unlike any human art I've seen, including cubist and other geometrical forms, they are pleasing to me and to many others. The computer-produced music I've heard has been haunting and weird and quite unusual, but also in a proper setting pleasing and even exciting. I'm told by computer professionals that when more computer people turn to this area, and when more artists and musicians learn programming or work with computer technicians, the range of computer art and music will be broadened almost without limit, adding new culture and pleasure to our lives.

New and different art forms are important and should be encouraged. But what will the computer do to existing cultural forms? Again, unless we recognize its probable impact and do something about it, the computer may have the very opposite and inhibiting effect, at least upon some forms of creative works. It is technically possible to introduce a literary work into a computer's memory bank, and thereafter have the computer review ("scan") it electronically and reproduce any part almost instantaneously. Some contend that such review does not infringe the copyright on the original work any more than does reading the book itself. For reference works such as Jane's volumes on ships, noninfringement could mean that the publisher's traditional market no longer exists. Who wants the whole expensive and bulky publication, when for a few cents and a telephone call the relevant information about the size, description, and location of the desired vessel is available? Dictionaries, encyclopedias, and other compilations of similar vein are all subject to similar treatment, which could reduce or even eliminate the author's and publisher's markets and therefore the financial incentive to produce the work in the first place.

Nor has society really begun to deal with the question of whether computer-produced art is in fact "art" or original creation; if so, is it to be given the same constitutional protection and encouragement as human-produced art? One would think that here at least the answer should be clear, for computer-produced art seems no more the creation of the computer than is sculpture the creation of the chisel. Computer hardware simply prints out whatever it is told to as the result of the software instructions it is given. And it is man who designs the software. Computer-produced art is thus analogous to a print produced by a wood block made by man, and should be given precisely the same treatment.

Although probably this will be the result when the problem is finally dealt with, even here there are no answers—only questions. In the single opinion in which it dealt directly with a computer-industry issue as such, the U.S. Supreme Court ruled that a particular kind of computer program was not covered by existing patent laws. Of course most art forms are covered by copyright, not patent, so that the case is distinguishable here. However, in language especially pertinent to the theme of this book, Mr.

Justice William O. Douglas, speaking for a unanimous Court (but of only six participating Justices), stated:

> If these programs are to be patentable, considerable problems are raised which only committees of Congress can manage, for broad powers of investigation are needed, including hearings which canvass the wide variety of views which those operating in this field entertain.

In late 1973 I conferred on this subject with the members of the Institute of Inventiveness and Protection of Intellectual Property at the Jagiellonian University in Cracow, Poland. The interest was so great that the session, planned for an hour and a half, went on for more than twice that time late into one evening, and then began again the next day. Everyone was quick to recognize the enormous implications of this electronic revolution to the special legal field in which they had chosen to specialize, and to conclude that substantial research and study were warranted.

Perhaps to others the art issue (and even the patent issues about which Mr. Justice Douglas was speaking) seem like unimportant problems, easily solved. Scanning could be made an infringement, subject to a license fee; the author could be rewarded for his contribution in any number of other ways as well. Computer-produced art, music, poetry, and literature could be given the same protection as any other form of artistic effort. If so, one wonders why even in this relatively simple area no one has acted so as to make clear the author's rights and thereby to encourage old and new art forms. Even here there is little or no appreciation of the computer's impact on our lives. Even here there is none of the understanding which would lead us to analyze and plan and thereby control what happens, so that we can assume intelligent responsibility for our futures.

MAN'S IMAGE OF MAN

Quite possibly at some future time the most significant ultimate impact of the computer—and the most difficult to articulate—will reveal itself to have been mankind's own "macro" and "micro" images of itself. The macro-view the computer gives us of what we are and what we mean may make us wonder whether we really are and what we mean may make us wonder whether we really are anything more than animals, whether there is in fact a God, and what is humanity; the micro-view it furnishes of ourselves as individuals may be very different from what we now perceive. If it is one we don't like, there may be devastating psychological consequences.

Copernicus' discovery that the earth is not the center of the universe and Darwin's theory that the ape is man's ancestor ultimately changed human life on this planet in profound ways. Even more dramatic consequences may ensue when the computer's ability to duplicate man's mental efforts is appreciated.

Can computers think? Paul Armer, one of America's most able and thoughtful computer professionals, perceptively answers tongue-in-cheek:

> No, because we keep redefining "thinking" as a process just a little bit beyond whatever we find the current ability of the computer to be.

Computers play chess and improve their games by learning from their mistakes. In fact, unlike humans, the well-programmed computer learns so well that it will not make the same mistake twice. Could one distinguish whether a respondent at the other end of a telephone line is a computer or a human, or is "thinking," by the nature of an unspoken but accurate answer? (Computer people call this "Turing's Test," after a paper on this subject published in 1950 by the English mathematician, A. M. Turing.)

Researchers at Stanford University in California have programmed a computer to simulate paranoia. When questioned over a terminal, the computer provides written answers that even psychiatrists are unable to distinguish from those of a person suffering from the condition. Eight psychiatrists were asked to question the persecuted computer as part of an experiment. One conversation went:

Doctor: What problem brought you to the hospital?
Computer: I am quite upset.
Doctor: Could you tell me why you have been upset?
Computer: People get on my nerves sometimes.

Doctor: How do they get on your nerves?
Computer: It bothers me when people stare at me.
Doctor: Why do people stare at you?
Computer: What about my looks?

The doctors were also asked to question actual patients via a computer keyboard. Then transcripts of all the interviews with patients and computers were sent to psychiatrists around the country who were asked to judge whether each interview was conducted with a patient or with the computer. Only 51 percent of the answers were correct. Similarly, two psychiatrists found that there were no significant differences between the answers given by subjects to questions posed by human therapists and computer therapists.

At one of the computer industry's conferences in 1969 or 1970, the audience saw an amazing and exciting film of travel through interplanetary space at the speed of light. A computer scientist had programmed abstruse and esoteric mathematical formulas into a computer (the lay equivalent would be the computer input of Einstein's theory of relativity). Then he had taken pictures of the computer's output showing how a road with telephone poles would look if one traveled down it at varying and increasing speeds. As one traveled and the speed increased, the telephone poles came closer together, then bent in peculiar fashion, and finally jumped all over the screen. Many of those to whom I later spoke in the audience shared my feeling that we had experienced something special and unique that would later somehow have major significance to our lives.

After the movie we were told of an experimental and not yet fully developed application of the same principles, in which eye pieces connected to a computers' output were placed over each eye, and the individual permitted to visualize how it would be to travel on the moon, or on a planet of high gravity, or under any other circumstances in which the rules for existence could be quantified and introduced into the computer. This may not be the cellular translocation to distant planets and stars which we see on television, but it does permit us to experience such existences. Even if sensing is not reality for all purposes, it can be for treatment of mental conditions and, more material here, for generating in man a new and different image of his own existence.

The human brain is an incredibly complex organ; it is doubtful whether within present lifetimes we shall ever be able to duplicate *all* its functions in a single machine, even if we want to. But it is at least arguable that the brain is made up of a massive number of computer-duplicatable parts, such as memory and processor, and that we will one day be able to replicate any specific process or function we wish to emulate. When we do, the chances are that the computer will reproduce the process more quickly, more efficiently, and with less prejudice and error than the human.

When mankind comes to appreciate that the computer can duplicate the thinking process which he had thought distinguished him from the animal, it will call for a reevaluation of himself and what he is in light of this new revelation. The consequences may be unpredictable, but they will surely be enormous.

Just as the computer can make us see humanity very differently in its relationship to God and the rest of the universe, on a more personal and intimate "micro" basis it can make each of us see ourselves as others see us, and not as we now think we are. The consequences can be even more shattering, for few of us really consider ourselves "bad" or our own conduct unjustified, whatever may be the contrary view the rest of society has of us.

My first vivid recollection of my experience 20 years ago as a federal criminal prosecutor is of an abortive effort to persuade a narcotics pusher to disclose his source of the drug. I offered him "maximum consideration" in sentencing—what is now called "plea bargaining." This man was the dregs of the social structure. His words, the substance of which still stand out in my memory 20 years later, were:

> Mr. Wessel, I don't know whether your father was rich or how you went to college and law school and got this job. But you've got your racket and I've got mine. Don't ask me to do anything wrong, and I won't ask you to.

Nothing could have made me see more clearly that my attempt to get him to betray his associates was as immoral and unethical to him as my accepting a bribe to fix the case would have been to me.

I recall also Joe Profaci, the "olive oil king" and then leader of the "Mafia," coming into court each day during the lengthy Apalachin criminal-conspiracy trial. He was a small, gentle, and kindly appearing old man with a pleasant face and manner. He would show me pictures of his grandchildren, and of himself with the Pope at the Vatican, with real tears in his eyes at the tragedy of the public trial and its harm to his family and reputation. To him at least, whatever he had done was the product of his own special environment and was well-justified by circumstance; everything was dictated by the rules by which he had been brought up and lived, and the clash with the laws of a different society did not make it morally wrong.

Indeed, given the subjective test of self-image, in 25 years of experience with people charged with crime, fraud, marital infidelity, misrepresentation, and innumerable other forms of misconduct, I can truthfully say that in one sense, "I have never met a bad criminal." Apart from those few whose self-images have been destroyed by mental breakdown, we all justify our own conduct and see ourselves as "good" by some test we accept. I suspect this would be true for anyone you chose, whether Attila the Hun, Josef Stalin, or Adolph Hitler, depending on your bias. More recently those who condoned violence and property destruction in protest over the Vietnam war policies, and those who justified criminal conduct in the Watergate episode, all believed that what they did was right in pursuit of *their* concepts of the national interest.

The computer threatens this self-image by giving us an intimate and objective view of our own motivation and conduct. A study underway of disparities by judges in sentencing criminal defendants may be illustrative. The differences in sentences meted out to similar defendants convicted of similar crimes by *different* judges even of the same court is almost a national scandal. Each judge can deal with and even criticize and pontificate about this, however, by simply considering the other judges "hard" or "soft" on criminals ("old turn-'em-loose Bruce"), or just plain wrong. But any experienced criminal-trial counsel, for prosecution or defense, knows of disparities in sentencing by the *same* judge which are similarly not based upon any proper distinguishing factor, or upon factors which the sentencing judge is willing to admit he uses, proper or no. Fortunately for the sentencing judge's self-image, the tremendous number of variables in sentencing (nature of the crime, age, prior criminal record, mitigating or aggravating circumstance, education, opportunity, and the like) renders it almost impossible to sense this kind of discrimination except by third-party objective intuition. Pragmatic prosecutors and defense counsel cannot suggest it to the judge himself. But trial lawyers pride themselves on knowing each judge's inner prejudices, on being able to select the right judge for the right case, and on their abilities to jockey procedurally to get before that judge. And prosecutors even bargain with defense counsel for the favor!

The computer, however, finally makes it possible to objectively analyze all these variables. Once analyzed for a significant number of defendants sentenced by the same judge over a period, the inference will become inescapable even to the judge himself that he discriminates in favor of black or white, rich or poor, alien or native, male or female, or what have you. When this unassailable computer demonstration of his own prejudice is shown to him, what then becomes of the judge's sense of his own integrity and impartiality and of his confidence in his ability to do justice later on?

Just as the computer may force the judge to see his own bias, it may compel each of us to see our own defects. The mental hospitals are full of people who do not like what they see when they look at themselves. One cannot even guess at the consequences when the computer lets us each take a closer look at our own selves. But we must be ready!

Once more the need for broad, deep public understanding of the computer and its consequences seems apparent. Not only technicians, educators, and legislators, but also ministers, psychiatrists, psychologists, and other spiritual and personal advisors must be made aware of the new reality in our lives; they must understand the pressure and help us deal with it. Indeed, the computer's substantial but still unpredictable impact upon human labor and intellectual effort, and even upon mental processes and religious and ethical concepts, requires that there be one final controlling principle for the computer age:

> **Computer Societal Impact:** *Government officials, professionals in and out of the computer industry, educators, and other leaders must study the impact of the computer on society, discuss and publish their efforts, and inform the public of their views.*

9 CONTROLLING THE COMPUTER

One of my most learned and sophisticated lawyer friends telephoned me one morning, asking, "What is 'SHAKE' "? "I don't know," I replied, "why ask me?" He answered that it was a term used by a computer client, and that he assumed I would know all about it. I said it sounded typical of the innumerable acronyms computer people like to use and suggested that he ask his client. This, however, he was most reluctant to do, and he asked me to try to find out from a computer colleague.

I called three computer professionals I knew, but none of them had heard of "SHAKE." Finally one called me back and said he had just found it listed in a new supplement of computer programs and that it was the acronym for a new program useful to a particular industry with which I knew the client was concerned. I called my friend back and he was delighted to now know what he was dealing with.

This little anecdote is not just an isolated sport, but typical of the reluctance most noncomputer people feel about revealing their ignorance of computer matters. As with many other esoteric new ideas and disciplines, people in general feel "spooked" by something different, and cover their ignorance by looking wise and saying nothing. This attitude helps explain why our nation's leading economists refrain from writing about or discussing the computer industry, although it has one of the most unusual structures of any major industry. It explains why sociologists, philosophers, and others who should long ago have been thinking and writing about the computer steer clear of the subject and thereby keep themselves and the public in ignorance. It explains why our universities, which should be the centers of intellectual research and writing about this new technology, have few courses and less written output dealing with computers and society.

Harvard's Program on Science and Technology, which was engaged partly in making a start in studying computer societal impact, closed up in 1972, and the National Academy of Sciences' Computer Board, which accomplished little but at least had a portfolio to proceed, was also placed in limbo in late 1972. Apart from the National Science Foundation's Computer Impact on Society Section, little new is being developed or proposed to take their places. The Nixon Administration's deemphasis of basic scientific research in general, and restrictions on funding these kinds of efforts, suggest that not much is on the horizon. The fact that some of this may be justified by economic priorities doesn't help.

The preceding chapter referred to Mr. Justice Douglas' comment about the need for information about computers, in the only case in which the U.S. Supreme Court dealt directly with a computer-industry issue as such. Even the procedural aspects of that case tell a great deal about the information vacuum concerning the computer industry—and even they have gone largely unnoticed in the computer industry itself.

Three of the Supreme Court justices in the program patent case did not participate in the decision, perhaps because they held IBM stock or had some other interest or prior connection. (IBM was one of the many organizations involved in the case.) The same three justices also refrained from participating in a subsequent case involving review of an interim procedural ruling in the U.S. government's antitrust case against IBM. The decision they were reviewing in the subsequent case was in turn from the Court of Appeals for the Second Circuit sitting *en banc* (that is, the full bench of active judges), where three of the eight active judges had also excused themselves from participating, presumptively for a similar reason. One wonders how many federal dis-

trict court or state court judges have had to recuse themselves from participating in the many private actions involving IBM over the years.

The fact that *one-third* of our U.S. Supreme Court and a higher fraction of an intermediate federal court will not sit on cases involving the computer industry certainly focuses on its unique economic structure and IBM's special position in it. (Perhaps it also says something about our judiciary.) The computer industry is one in which even giants from other industries such as GE and RCA simply cannot make it and are forced to withdraw from the hardware mainframe (central portion) manufacturing segment. It is an industry in which major computer manufacturers themselves, such as National Cash Register and Control Data Corporation, have to combine parts of their operations in order to be able to compete. It is an industry in which the U.S. Justice Department has launched a major antitrust attack (against IBM) charging monopolization. Yet it is also an industry which the same Justice Department terms "highly competitive" in the peripherals (input and output devices and the like), software, services, and other segments.

There is so little available information about the computer industry that in one recent private antitrust suit a federal court even found it necessary to authorize the depositions of an unprecedented 3300 third parties. Answers to written questions were required by court order, in an effort to find out something about the composition and structure of the industry. Many of these third parties complained bitterly of the imposition on their affairs and the possible harmful disclosure of confidential trade secrets. In other industries there are all kinds of publicly available ongoing statistical analyses, treatises, texts, and the like, furnishing the kind of information sought in this case. The unprecedented judicial action could be justified only because of the unprecedented absence of this usual kind of public information. That even this action was not sufficient is indicated by the fact that a similar new but more intensive and far-reaching industry survey is already underway in another case.

The inevitable consequence of this lack of information and fundamental research and analysis is that we are forced to use outmoded and inadequate concepts and tools to deal with computer-age problems, sometimes with almost ridiculous results. The present brouhaha about the taxation of computer software is a good example of this computer-industry anachronism.

Computer software includes the instructions which enable the computer's physical equipment (hardware) to do the work. Without it, most computer hardware (a few special hardwired computers excepted) is just a mass of inanimate wire and metal; with software, the machine comes to life and begins to produce its output. Computer software is a multibillion-dollar industry; when all its aspects are broken out of the manufacturers' integrated operations ("unbundled"), it may well exceed 10 billion dollars annually and be larger than hardware. It is growing more rapidly than hardware; it is software in which future computer breakthrough is most likely to occur, permitting communication between man and computer in something much more closely approximating ordinary written language than is commercially practical at present, and simplifying oral and visual communication as well. With such a valuable property there is little wonder that financially pressed state and local government authorities are beginning to give attention to its taxation.

One key current legal question has been the applicability of the California tangible personal property tax to software. The obvious resulting debates are whether software is tangible or intangible; if so, what is its "situs" or location, for purposes of imposition of the tax at the moment of assessment (usually a specific date and time); who "owns" it, for purposes of determining who pays the tax; and what is its "value," for purposes of deciding how much the tax should be.

This is logical enough *until* one begins to appreciate just how very different software is from any other property right we have ever known, and how nonsensical it is to try to apply concepts applicable only to a very different world. For software is an ubiquitous thing whose use is often divided among so many different units as to make the old concept of "title" or "ownership" almost meaningless, with a value sometimes so uncertain as to be illusory, and a form which can be changed and moved at a moment's notice at almost the speed of light.

Consider, for example, a typical time-sharing application. In "time sharing" the operating portions of a computer's CPU (central processing unit, the calculating part of a computer) are divided into a large number of tiny time frames of only milliseconds or even microseconds (thousandths or millionths of a second) in duration. These time frames are rotated among large numbers of remote customers who interrogate and receive responses through terminals which are much like LEXIS typewriters with television screens. The terminals are connected to the computer by telephone lines. Even though each user is sharing the computer with many others, the input and output devices geared to human and mechanical motion are so slow compared to the electronic speed of the CPU that the computer's responses seem practically instantaneous. Each user acts as though he has the computer all to himself.

In one such existing commercial time-sharing system, customers all over the world can interact with the vendor's central computers at their pleasures. Each has the right to use the vendor's software and, in some cases, to license its use to others. They don't know or care where the physical computers are located, much less where the software may be at any given moment. If California imposes a tax on the software, the vendor can move his computers to Nevada, or Mexico, or Canada, or outside the three-mile national boundary. If the tax is imposed at 12:01 A.M. each April 1, for example, he can erase all tapes, discs, drums, or other forms of the product the day before, store the software in another state (or even conceivably in an orbiting satellite), and reproduce it the next day in the identical earlier form; if "value" is based inversely upon number of copies, as is common with some other products, he can reproduce a million in an instant. Almost any other traditional test or analysis is subject to similar defects. Indeed, if the software problem becomes too great, the instructions can be stored in semipermanent memory in the computer itself. (This is called "firmware," a product which has characteristics of both hardware and software.)

All of this is not to contend that software should not be taxed, but to emphasize the great need for public consideration and discussion of the underlying economic and social considerations involved in dealing with this new and quite unique creature. In light of its special characteristics, does it really make sense for one state to impose a personal property tax on the software in a legal research system serving attorneys nationally or worldwide, simply because the hardware is located within its borders? Would such a tax constitute an unreasonable burden on interstate or foreign commerce? Might it persuade the proprietor to move his hardware elsewhere, even if it would otherwise be uneconomical to do so, or to buy firmware where software would be more useful? Shouldn't the amount of the tax be determined on the basis of considerations such as the fairness of the tax burden in comparison to that imposed on other properties and activity, and the economic and social benefits to be derived from encouraging the computer industry?

Other computer-oriented legal issues are similarly considered in ancient terms that are inappropriate for the unique problems and opportunities of the computer age. Whether a lien exists on a computer tape, or other device to which information has been added, may be determined on the basis of whether the electronic impulses constitute an "improvement" of the tape—without an examination of the interplay of supplier-user rights and interests. Whether the Robinson-Patman Act's prohibition against discriminatory pricing (an anathema to many antitrust lawyers in any event) applies to computer output may be decided by whether such output is a "commodity" or a "service," rather than determined by basic economic considerations. Indeed, in the intellectual property area the niceties seem almost deliberately contrived to avoid the fundamental question of the extent to which these new forms of effort deserve encouragement. Thus the issue of patent protection for computer software may turn on whether software is technically classified as a "machine"; the issue of whether computer-produced music, poetry, or art is protectable may turn on whether the computer or the programmer is the "author"; the issue of whether a computer *reproduction* of a copyrighted work (or even the *introduction* of such a work into the computer's data bank) constitutes infringement may turn on whether the computer is "copying" and whether a "copy" has been made.

It doesn't make sense to predicate significant consequences on the outcome of debates regarding largely irrelevant concepts developed in different times for different purposes. Software is precisely the same thing whether it is classified as "tangible" or "intangible." The computer is no different if it is an "author." The current exchanges are much like arguing about how many angels can dance on the head of a pin. It doesn't make any practical difference!

Dr. Bruce Gilchrist, another of our nation's leading computer professionals, and I published a computer-industry study in 1972. It pointed out that this lack of information and understanding about the computer was characteristic of federal government operations. This results in federal agencies often taking actions in ignorance. Many important regulatory actions are completely inconsistent with each other, to the detriment of society at large as well as the computer industry. Even the government's antitrust prosecution of IBM, by far its most important pending antitrust case and perhaps the most significant ever brought, has been a Kafkaesque effort (in January, 1974 it began its sixth year of "indefinite postponement"—although there is finally hope that trial at least will begin in late 1974); thus far it has failed to appreciate the need promptly to resolve the charges and defenses one way or the other, so that we can move forward with the work at hand.

I do not suggest that action by the government in its antitrust prosecution and research and debate in other computer areas will furnish *all* the answers to *all* the issues. Undoubtedly there are some computer-societal effects which we simply are not equipped intellectually to anticipate. MIT Professor Joseph Weizenbaum, who has thought long and hard with regard to computer societal issues, argues that the invention of the microscope in the middle of the seventeenth century enabled man to see microorganisms and paved the way for the germ theory of disease. Until then the dominant theory was that disease was a punishment visited by God on man, with the cure being penance and the balancing (by bleeding, for example) of body "humors" which had been brought into disequilibrium in accordance with divine justice. The discovery of extremely small living organisms contributed to the idea of a continuous chain of life which was a necessary intellectual precondition for the emergence of Darwinism. The germ theory of disease and the theory of evolution altered man's conception of his relationship with God. With other developments, these theories helped contribute to the diminution of the power of the church and, more generally, to the questioning of the basis of hitherto unchallenged authority. Surely the seventeenth-century observer of the invention of the microscope could not possibly have anticipated all this. For much the same reason it would take a degree of conceit beyond arrogance to believe that we can foresee all the future effects of the computer. But the fact that we cannot anticipate and provide for *everything* should not stop us from doing as much as we can. The 1974 Cable Committee report put it this way:

> We can no longer permit technological innovation to "just happen" and then attempt to "regulate away" the adverse effects.

It bears repetition that our country's intellectual and political leadership, with more than enough other business to transact, chooses to remain silent rather than speak, in my opinion largely because of fear of displaying ignorance. The computer profession must bear at least part of the responsibility for this, because it has needlessly but almost deliberately developed an esoteric language all its own, which others must struggle to understand. As one trial judge wrote after eight full days of listening to evidence in a computer case:

> After hearing the evidence in this case the first finding the court is constrained to make is that, in the computer age, lawyers and courts need no longer feel ashamed or even sensitive about the charge, often made, that they confuse the issue by resorting to legal "jargon," law Latin or Norman French. By comparison, the misnomers and industrial shorthand of the computer world make the most esoteric legal writing seem as clear and lucid as the Ten Commandments or the Gettysburg Address; and to add to this Babel, the experts in the computer field, while using exactly the same words, uniformly disagree as to precisely what they mean.

With too few exceptions, computer people attend to their own affairs and their own problems. They are the technological experts who are building our New World, yet who are in important respects almost apart from the rest of lay society. Their judgments, guidance, and advice are essential if society is to be able to deal with what they are developing. Two rules for conduct are therefore necessary to make public understanding of the computer a possibility:

> **Public Understanding Rule 1:** *Laymen must not hesitate to ask questions of computer professionals because they consider the computer too complex, or are reluctant to disclose their ignorance.*

> **Public Understanding Rule 2:** *Computer professionals must answer lay questions in terms which are understandable to laymen.*

A glimpse at the exciting possibilities to be achieved by tapping and interacting with the talents of computer scientists was provided when the first truly interdisciplinary conference of computer scientists and law faculty members concerned with computer-related problems was held at the Stanford campus at Palo Alto, California in June, 1973. The two-and-one-half-day meeting was cosponsored by the country's amalgam of computer professionals, the American Federation of Information Processing Societies (AFIPS), and Stanford Law School; it was financially supported in part by the National Science Foundation and IBM. From before breakfast until late into the evenings, a score of computer-science leaders explained computer technology, while law teachers identified the legal considerations involved. In an effective synergistic relationship, a host of computer-societal issues were identified and explored during the formal conference meetings and in the many informal off-hour "rap" sessions. They included not only those discussed earlier in this book, but also other issues incident to any similar major scientific or technological advance, such as the extent to which the computer will replace or displace labor, or the effects of reliance on computer

computation and analysis on human intellectual capacity and effort.

Although of course this conference was only the most rudimentary beginning to the broad, in-depth, long-term effort required, it furnished at least a hint to those present of just how much can be done if only the communications barriers between computer scientists and the rest of society are broken down, and if an exchange of information and ideas concerning computer-societal problems is developed and participated in by *all* the relevant disciplines (sociology, economics, psychiatry and psychology, law, the creative arts, and so on) as well as by the public at large.

Our country's intellectual leadership has a responsibility to inform and advise, so that the people may decide. It is not carrying out that responsibility. We must insist that it do so. The rewards are too great and the dangers of any contrary course too serious to permit otherwise.

EPILOGUE

The turn of the year 1973–4 was the time of the Arab/Middle-East oil embargo and growing awareness of the "energy crisis."

In the mid-1950's I was trial counsel to Standard Oil Company of California in two huge government litigations involving Middle-East petroleum. The first sought to recover alleged overcharges under the Marshall Plan totalling some 66 million dollars; the second was an antitrust attack against the so-called "international oil cartel."

One of the standing jokes among the many counsel in the litigations was that both cases were based on a gigantic fraud perpetrated by the oil industry, which charged money for products which didn't exist—there really was no such thing as "oil" at all. The "proof" cited in support of the charge was that no one ever saw any oil company product from the time it allegedly was taken from the ground in Qatar, Kuwait, or Saudi Arabia, until the time it was used up in an automobile's gas tank. The dials, gauges, meters, and other evidences of activity were equally spurious. The oil companies allegedly concealed their conspiracy by claiming that the nonexistent petroleum products were in pipelines, tankers, refineries, pumps, and the like, where they could not be seen; every so often an odor was permitted to escape or a few drops were deliberately allowed to spill out of a gas hose to lull the public into a sense of security. The case would be broken some day when a brilliant government investigator looked into a hatch in an apparently full oil tanker and found it empty.

The alleged fraud was a put-on, of course. But good humor usually contains a kernel of fact. That fact was that the petroleum business was so highly automated—even at that early time—that no one ever saw or touched its products. Today computers read meters, make production adjustments, control the mix among the many complex products of the refinery process, schedule home heating oil deliveries, and have otherwise made possible further and now almost total automation of the petroleum industry. Without computers there would be an energy "catastrophe," not just a crisis, for there would be almost no petroleum products at all.

These pages have tried to show how, at an increasingly accelerated pace, more and more social and economic as well as production aspects of our modern industrial society are and will be similarly dependent on the computer for continued effective functioning. Some computer scientists speak of the computer as nothing more than an incredibly high-speed calculating machine, made up of bits and pieces of metal, wire, plastic, crystal, and electricity. In the sense in which they use the term "calculating machine," they may be right. But in a societal sense, the computer is much more than a machine.

Telegraph, telephone, radio, and television are all just ways of communicating quickly. But the very capacity to communicate with electronic speed has changed the fabric of life on this planet.

The steam engine, the automobile, the propeller and jet plane, and the rocket are just fast ways of moving about, but the speed with which we move has also changed our way of life, for better or for worse. We no longer transact business or deal with each other as we did before the jet age.

It took a relatively long time for wireless communication and air travel to work their changes, and there was opportunity to begin to create new institutions and ways to deal with what was happening. Whether or not those institutions are good ones or are adequate to deal with the new ways of life we enjoy is an open question, the answers to which are not in yet. But the

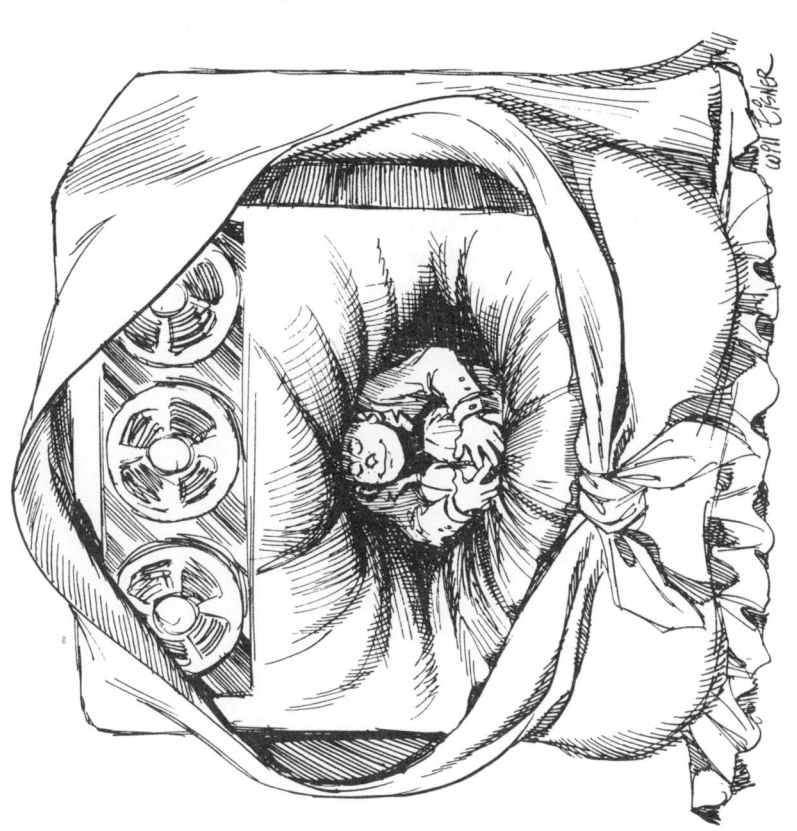

opportunity did exist, and we had our chance, good or bad. We may still have it.

The pace of computer life is accelerated, however, and we will not have the same length of time in which to deal with the changes created by the computer. The computer age is here. Just as astronauts in orbit cannot return to planet Earth without the computer, we have already reached the point of no return. The computer dominates major parts of our lives and will inevitably grow in its impact. Rather than shudder, tremble, and wring our hands in dismay, let us accept the computer and deal with it just as we do any other tool we develop.

We are in urgent need of public discussion, debate, and understanding of the new technology. Let us not permit anyone to tell us that any socially relevant aspect of the computer is too complicated for us to understand. Let us revel in our ignorance, demand explanations of the unclear, and delight in our acquired ability to handle this machine and make it work for us.

APPENDIX

THE TEN COMMANDMENTS OF COMPUTER USAGE

These commandments express principles of computer usage only in the broadest general terms. We do not yet have the information to do more. Recognition of fundamental concepts of this kind, however, is an essential first step to later defining such details as what is a "mass" data bank, or what is "reasonable" under varying circumstances. Unfortunately, except among a very small concerned segment of society primarily within the computer profession itself, the process of developing these necessary principles of conduct has not yet begun.

The Computer Utility

	Page
First Commandment. Access to a computer utility system shall not unreasonably be withheld.	13
Second Commandment. The information disclosed by a computer utility system seeking response must be such as to permit the respondent to provide an intelligent answer.	14
Third Commandment. The information furnished by a com-	14

puter utility system must be such as to serve the public interest.

Fourth Commandment. A computer utility credit card shall not unreasonably be withheld from any individual.   24

The Data Bank

Fifth Commandment. A mass data bank shall be permitted to operate only if the benefits associated with its operation outweigh the related risks.   56

Standards

Sixth Commandment. Computer standards should be fixed by fairly selected and representative public organizations, so as to encourage maximum reasonable interchange among computer systems and between economic units, without unreasonably impeding technological development.   81

Public Computer Services

Seventh Commandment. Public and quasipublic-sponsored computer services must be supplied on terms and conditions which result in their fair and equitable distribution to the public.   90

The Computer Economic Grid

Eighth Commandment. The failure of a discrete unit of a computer economic grid must result in immediate disconnect from the grid without unreasonable harm to or interference with the rest of the system.   94

Human Response

Ninth Commandment. The supplier of computer services to the public must afford the ultimate consumer reasonable human response and interaction, or be liable absolutely for error and harm done.   104

Computer Societal Impact

Tenth Commandment. Government officials, professionals in and out of the computer industry, educators, and other leaders must study the impact of the computer on society, discuss and publish their efforts, and inform the public of their views.   118

    Public Understanding Rule 1. Laymen must not hesitate to ask questions of computer professionals because they consider the computer too complex, or are reluctant to disclose their ignorance.   128

    Public Understanding Rule 2. Computer professionals must answer lay questions in terms which are understandable to laymen.   128